D1085334

TECHNIQUES OF SAFETY MANAGEMENT

TECHNIQUES OF SAFETY MANAGEMENT

DAN PETERSEN

Safety Management Consultant
Colorado State University

Second Edition

McGRAW-HILL BOOK COMPANY

New York St. Louis San Francisco Auckland Bogotá Düsseldorf Johannesburg London Madrid Mexico Montreal New Delhi Panama Paris São Paulo Singapore Sydney Tokyo Toronto

Library of Congress Cataloging in Publication Data

Petersen, Dan.
Techniques of safety management.

Includes bibliographies and index.
1. Industrial safety. I. Title.
T55.P37 1978 614.8'52 77-9384
ISBN 0-07-049596-3

1234567890 KPKP 7654321098

The editors for this book were Jeremy Robinson and Esther Gelatt, the designer was Elliot Epstein, and the production supervisor was Frank P. Bellantoni. It was set in Baskerville by University Graphics, Inc.

Printed and bound by The Kingsport Press.

To Nadyne

CONTENTS

PREFACE

Since the first edition of this book was published, in 1971, a great deal has happened in safety management. The most publicized change was the advent of OSHA, the Occupational Safety and Health Act, which went into effect shortly after the first edition appeared. The most important change, which occurred in spite of OSHA, has been the "growing up" of safety. OSHA emphasized a refocusing of attention on physical conditions, and for a while we in the safety profession followed in this direction. Perhaps our refusal to continue focusing on conditions indicates the maturation of safety management. We know that things do not cause accidents and we have refused to let the passage of a federal law direct us back into nonproductive activities that have proved ineffective in the past.

In recent years we have made tremendous strides toward professionalism. We have defined our scope and functions as professionals and we are now being recognized as certified safety professionals (CSPs) and as registered professional engineers in safety (PEs). Curricula for baccalaureate and graduate degrees in safety have been established, and many people in this country now hold master's degrees and doctorates in the field. Not long ago only a handful of schools taught safety; now many do. We have also developed principles to guide us. This is important, for without principles—without our own body of knowledge—we cannot consider ourselves professionals.

Another indication of our maturation is the development and open discussion of different philosophies of, and approaches to, the practice of safety management. We used to be able to talk only about a traditional approach, based on the "three E's" of education, engineering, and enforcement. We now discuss total loss-control approaches, operational-error approaches, systems safety approaches, psychological approaches, and sociological approaches.

A book first published in 1971 must undergo major changes if it is to fit today's world. This edition incorporates many revisions. Part 1 has been changed to reflect the enactment of OSHA, whose existence we must recognize. The book deals with OSHA somewhat lightly, however, and treats it as a subject separate and distinct from safety management. We in the safety profession must be concerned about both compliance with OSHA and accident control; they are not the same thing.

Parts 2 and 3 have undergone major changes. Instead of three chapters on managing and measuring safety performance, there are now eight chapters on these important subjects. When we learn to manage and measure line safety performance properly, we begin to get results. Part 2 deals with management's direction and accountability, and the five chapters in Part 3 are devoted to measuring safety performance at different levels in the organization.

This edition of the book reflects the fundamental changes in our knowledge and beliefs that have taken place in recent years. The first edition preached line accountability as the key to safety performance, and this edition does the same. However, we know more now about measurement, and since it is the key to accountability, it must be discussed in more detail. Effective measurement of the line organization's safety performance is the most important item in the safety program. With effective measurement you get motivation, performance, and results. Without effective measurement, you get none of these.

Part 4, which has also been changed and updated, deals with motivation at various levels in the organization. Part 5 discusses a number of additional safety techniques, some of which were covered in the first edition and some of which are new. Chapters on utilization of the computer and on profiling have also been added. For my discussion of computer techniques, I am indebted to Bill Pope, president of Safety Management Information Systems, Inc., one of the true pioneers in safety management. For the ideas on profiling, I wish to thank Jim Tye, director general of the British Safety Council, and Bunny Matthysen, managing director of the National Occupational Safety Association of South Africa.

The ideas put forth in this book are not original. Ideas rarely are—they are transmitted from person to person, and they grow in the process. This book attempts merely to record current ideas and assess their value. The thoughts expressed come from many sources and are the product of years of collecting, developing, polishing, teaching, and managing. Certain individuals, however, have contributed heavily to the material presented in the book.

D. A. Weaver, director of Policyholder Educational Services for Employers Insurance of Wausau, helped develop many—perhaps most—of the ideas in this book. He and I worked closely for many years. Specifically, his contributions can be found in Chapters 1, 2, 3, 4, 14, and 16. He coauthored portions of Chapter 3 and he invented the tracing system described in Chapter 14 and wrote most of that chapter.

Dr. Phil Brereton, director of executive and professional development at Illinois State University, also helped develop some of the ideas in the book. Ray Campbell, former safety analyst for Industrial Indemnity Company, did most of the work on the special-emphasis program described in Appendix C. Ray, with the help of Norm Leaper, of Industrial Indemnity, wrote the manual for the campaign described in the appendix and worked on the concepts in the campaign. Bob Sasser, then of Employers Insurance of Wausau, wrote portions of Part 4.

The people mentioned above were paramount in the development of this book. My thanks go to them and to all the individuals and companies whose materials are used throughout.

DAN PETERSEN

INTRODUCTION

The intent of this book is to fill some voids left by other safety textbooks and references, to challenge outdated theories on which we sometimes rely, to spotlight new techniques that are available but not often used, and to offer outlines for work in areas we often evade. This text is not a comprehensive reference book for the student or professional; rather, it is intended to challenge thinking. It seeks to add to (and in some cases question) the basic existing textbooks in accident prevention. The reader who needs basic technical safety information will find that hundreds of sources supply it. Once a need has been defined, information can be found. This book has not attempted to repeat all this excellent information.

Techniques of Safety Management is intended for the practicing safety professional. Its purpose is not to teach basic safety, but rather to describe areas of weakness in the safety profession and to point some directions for tomorrow. It is directed primarily to the safety professional—whether full or part time—in industry, in government, and in insurance. It also is slanted toward the line manager in business who has a particular interest in loss control. Finally, it seeks to add to the information presented in the basic textbooks on safety engineering.

THE SCOPE OF THIS BOOK

The book consists of five parts and the appendixes.

Part 1. Safety Concepts

This part deals with beliefs concerning safety and is included in the book because of my firm conviction that safety needs a better conceptual framework than it now seems to have.

Part 2. Managing Safety Performance

It is the intent of Part 2 to show how the theory of Part 1 can be applied in practice. The practical application is discussed in terms of what management and the safety professional can do to accomplish loss control—what procedures and policies management can set up to accomplish its defined goals in safety.

Part 3. Measuring Safety Performance

Part 3 examines current ideas concerning the measurement of safety performance. This new part has been added to the second edition because of past deficiencies in effectively measuring line safety performance. Measurement is the key to line performance and safety specialists have been notably weak in devising measures that motivate—or even measures that work.

Chapter 5 sets the stage by examining criteria for good measures of performance. It attempts to answer the question, "What makes a measure good?" Then, bearing in mind the criteria selected, we look at various levels in the organization and suggest some measures for each: for the first-line supervisor in Chapter 6, for middle and upper managers in Chapter 7, and for the company as a whole in Chapter 8. Chapter 9 discusses measurements of national safety performance currently in use and suggests new ones. Finally, Chapter 10 examines the relationship between measurement and motivation.

Part 4. Motivating Safety Performance

This part attempts to relate some of the recent studies in motivation to accident prevention.

Part 5. Additional Safety Techniques

Part 5 looks at some newly developed techniques in safety (Chapters 14 to 17) and at the two potentially most costly areas of loss in most industries (Chapters 18 and 19), since the scope of safety includes more than control of injuries to employees on the job.

The Appendixes

The appendixes present examples of the principles discussed in Parts 1 to 5.

This, then, is the scope of the book. The five parts and the appendixes might also be titled: (1) Principles, (2) Application of Principles, (3) Measurement Techniques, (4) Application of New Theory to Safety, (5) Application of Principles beyond Employee Injury, and (6) Aids. The primary intended audience for this book is the full- or part-time safety professional in industry, private business, government, or insurance. The secondary audience is the student of industrial safety.

THE STUDENT

I would suggest that the student—whether studying on the undergraduate or graduate level or just starting in industry but new to the safety profession—use this book in addition to other texts, not as a basic text. The basic texts offer the technical knowledge that any safety professional needs; this book offers an additional challenge to thought.

THE SAFETY PROFESSIONAL

In most cases this person did not choose the profession, but was chosen by it. In industry it is common for management to select its safety specialist not for abilities and knowledge in the field of safety, but rather because of demonstrated abilities in some other field. One day the future safety professional is tapped on the shoulder by the boss, and the job is assigned, either as a full-time position in management or in addition to other duties. The new specialist is usually then in the position of having to learn an entire profession overnight, and without outside help.

The Industry Specialist

It is extremely difficult for the industry specialist starting out in this field. There are few good orientation "schools," although some have been established in recent years. Information about them is available from the National Safety Council, the American Management Association, insurance carriers, governmental agencies, etc. The specialist-to-be ought to take advantage of what is available. However, some courses are far better than others, and the new safety specialist should look at course content critically and should also check with safety professionals in the area before selecting a course. They might provide valuable information about available courses.

Even with such assistance as this, the position is largely self-learned and self-defined. This book, directed to the experienced safety professional as well as the neophyte, attempts to provide some help. It has been said over and over that "there is nothing new in safety—we just keep saying the same old things in new ways." This is foolishness—there is more new in safety right now than any one of us can know or comprehend—we have only to look to find it.

In recent years exciting things have happened in safety research and theory. Most of these new developments have not yet been effectively used. This is the job of the experienced safety professional in industry today, a job not performed well at present—to know and comprehend all the new developments in safety.

The Part-Time Safety Specialist in Industry

Most part-time safety people make safety the smaller part of their job. They keep up with new developments in the portion of their work that they consider to be the main part, but they do not keep up in safety.

This book is directed to part-time safety specialists as well as to full-time professionals. Part-time specialists need to modernize their thinking even more than professionals.

The Line Manager in Industry

Although this book is not aimed at line managers, I hope many of them read it, as well as people in top-management positions. The responsibility—the actual carrying out of all safety work—is a line job. Parts 2, 3, and 4 ought to be of particular interest to line and top management.

The Private Consultant

The consultant has a well-defined role, that of adviser to industry management. The consultant analyzes the customer's problems and suggests controls and therefore must be even more up-to-date than the customer. This book is intended for the consultant also.

The Insurance Safety Engineer

The role of the insurance safety engineer or consultant (or whatever the title may be) is sometimes not as clear as the private consultant's role. Insurance safety engineers should perform the same role as private consultants. Some do; some do not. Some insurance engineers only inspect—some of their customers prefer it this way. Most insurance engineers are competent and well prepared to advise and assist industrial management in its job of loss control. Today insurance engineers themselves choose the level of safety service that they provide—as consultants, engineers, or inspectors. Each engineer ought to determine on what level he or she presently is operating by honestly answering each of the following questions:

1. How am I measured?

    a. By number of recommendations

    b. By number of calls

    c. By results achieved with the customer

2. What do I leave with my clients?
   - *a.* Lists of things that are wrong
   - *b.* Lists of solutions
   - *c.* The results of my analysis
3. How do I schedule my contacts with customers?
   - *a.* Regularly—routinely
   - *b.* On the basis of need
   - *c.* According to current projects
4. What is my means of communication?
   - *a.* A recommendation pad
   - *b.* A letter to the customer
   - *c.* A formal report to management
5. Who is my customer contact?
   - *a.* The plant engineer
   - *b.* The plant superintendent
   - *c.* The manufacturing vice president
6. What do my customers expect from me?
   - *a.* Trouble
   - *b.* Help
   - *c.* Advice
7. If my customer has a poor loss record, what do I do?
   - *a.* Inspect more frequently
   - *b.* Call more often
   - *c.* Define the problems
8. If my customer has an excellent loss record, what do I do?
   - *a.* Inspect less frequently
   - *b.* Call less often
   - *c.* Analyze why

9. What do I believe causes accidents?
   a. Conditions
   b. People
   c. Systems failures
10. What is the key to safety?
   a. Inspection
   b. Employee participation
   c. Management direction
11. Who is the "key person"?
   a. The employee
   b. The supervisor
   c. Management
12. What is the purpose of inspection?
   a. To unearth hazards
   b. To remove hazards
   c. To find symptoms of problems
13. Which is best?
   a. Safety first
   b. Production with safety
   c. Safe production
14. If we reduce frequency, will severity also be reduced?
   a. Always
   b. Usually
   c. The causes may be different
15. What is the single greatest factor we must work on?
   a. Improvement of physical conditions
   b. Line acceptance of responsibility for safety
   c. Better procedures to fix accountability for safety

Each of the above 15 questions has three possible answers. In each instance, answer *a* is typical of the inspector, answer *b* is typical of the engineer, and answer *c* is typical of the consultant. Scoring one, two, or three points for each answer as indicated, if you scored 15 to 30, you are behind the times by 30 years; if you scored 30 to 40, you are behind 10 years. A score of 45 says you are up-to-date.

The questions give some insight into the subject matter presented in the following pages. This book is slanted toward the insurance engineer, as much as it is toward staff safety specialists in industry.

The Government Inspector

One of OSHA's primary problems concerns the function of compliance officers: they have been pure inspectors in the past. This will change as their role enlarges to include helping organizations as well as enforcing the law. As the inspector evolves into an engineer and a consultant, the job will be subject to the criteria that apply to the insurance engineer. I would urge the government safety person also to answer the 15 questions above in grading his or her service to assigned companies.

As this revision is being written, several governmental enforcement agencies are discussing other ways to help American business achieve safety standards. Some excellent approaches and ideas are being considered. Often the laws, as written, do not allow such help. Our laws, written for the worker's safety, often hinder achievement of their avowed goals. It is hoped that this will change. If government *can* and *will* help companies—through better methods, better approaches, and better people—the worker will benefit, as will the company.

The purpose of this book, then, is to fill some voids in other safety references and to challenge outdated concepts. It will aid the safety professional—whether in industry, insurance, or government—and the student, as an addition to other safety textbooks.

PART ONE

SAFETY CONCEPTS

Our present framework of thinking in safety should be examined and perhaps challenged more than it is. Much of what we do today is based upon principles developed long ago.

It may now be time to reexamine those principles and look at some newer ones that have come upon the safety scene. This is the intent of Part 1.

1

THE PRESENT FRAMEWORK

In this initial part of the book I would like to do two things: (1) briefly examine where we are today in industrial safety and (2) suggest a framework of thinking for tomorrow, a framework based in part on today's theory and in part on newer, more forward-looking theory. Let us first look briefly at where we seem to be—where the past has led us—by examining industrial safety during two separate eras—before and after 1911.

BEFORE 1911

Progress in industrial safety before 1911 was practically nonexistent. With no workmen's compensation laws, all states handled industrial injuries under the "common law." Under this common law the legal defenses available to management in industry almost ensured that it would not have to pay for any accidents occurring on the job. Under this common-law system employees did not automatically receive payments when injured on the job, as they do today. Before workmen's compensation legislation, the injured employee had to sue the employer for recompense.

When the employee did sue, the employer had four legal defenses. If the employer could show any of the following, the company would not have to pay the injured employee for the injuries suffered.

1. The employee contributed to the cause of the accident.
2. Another employee contributed to the cause of the accident.
3. The employee knew of the hazards involved in the accident before the injury was sustained and still agreed to work in the condition for pay.
4. There was no employer negligence.

In 1908 the state of New York passed the first workmen's compensation law, which said, in effect, that regardless of fault, management would pay for injuries occurring on the job. Before this, obviously, most accidents were handled under one of the common-law defenses, and hence management did not often have to pay for injuries resulting from accidents.

The New York law was held to be unconstitutional in New York; a similar law was passed in Wisconsin in 1911 and was held to be constitutional. This Wisconsin law set the stage for all the other states to provide similar laws. All states did, the last being enacted in 1947. Actually a New Jersey law was passed before the Wisconsin law, but Wisconsin's went into effect a few months earlier.

AFTER 1911

When management found itself in the position, by legislation, of having to pay for injuries on the job, it decided that preventing accidents from happening would be financially sounder than paying for them. This decision by industry gave birth to the organized safety movement.

Workmen's compensation legislation provided the financial atmosphere for industrial safety. Without the legislation the safety movement would be far behind where it is today. Workmen's compensation laws, in effect, state that regardless of fault, the injured employee will be compensated for injuries that occur on the job. Liberal interpretation of these laws in many states today goes even beyond this original intent.

Improving Working Conditions

In the early years of the safety movement, management concentrated heavily, if not entirely, on correcting hazardous physical conditions. This produced remarkable results during the first 20 years. The number of deaths due to accidents decreased from an estimated 18,000 to 21,000 in 1912 to about 14,500 in 1933. The death *rate* (deaths per million worker-hours) for that period indicates even better results. This reduction in the number of deaths was brought about largely by cleaning up working conditions. This was the first step, possibly because conditions were so obviously bad and possibly because people believed that these conditions were actually the cause of injuries.

Heinrich's Safety Philosophy

In 1931 the first edition of H. W. Heinrich's book, *Industrial Accident Prevention,* was published. This text in industrial safety was revolutionary, for in it Heinrich suggested that unsafe acts are the cause of a high percentage of accidents—that people cause far more accidents than unsafe conditions do. Heinrich's ideas were a departure from the safety thinking of the time. What he said, however, made sense to people in the field of safety, and his ideas were accepted. They were accepted so completely that even today we work largely within his framework. His work set the stage, in effect, for all safety work since 1931.

Perhaps it was because Heinrich proposed a philosophy for safety that his work was so important. Before the publication of his book, safety had no organized framework of thinking. It had been a hodgepodge of ideas. Heinrich brought them all together and defined some excellent principles out of previous uncertain practices.

Progress after 1931

As a result, safety progressed markedly after 1931. Accident frequency rates, according to the National Safety Council, dropped from 15.12 accidents per million worker-hours in 1931 to a low of 5.99 in 1961. It was back up to 13.10 in 1975. The national severity rate dropped from 1,590 days lost per million worker-hours in 1931 to a low of 611 in 1971 and was at 752 in 1975. These indicators told the successful story. Safety people were proud of their accomplishments—as indeed they should have been. Basically, they were able to achieve so much because they did what Heinrich had said to do. He had produced a formula that worked.

SAFETY TODAY

Today we safety professionals are still proud of our progress since 1911, or even since 1931. We still quote the frequency- and severity-rate reductions we have accomplished since those early years.

But do such comparisons give us the true picture of our accomplishments? Let us think in terms of the period from 1961 to 1975 instead of from 1931 to 1975. The figures on page 15 show the true story. If we can believe a combined rate figure, we have made absolutely no progress in safety since 1961. In fact, we have lost ground. We are definitely regressing today, and the questions before all safety professionals are, "Why has our progress ceased?" and "How can we start the rates going downward again?" We must ask ourselves these questions now, for we seem to have lost our momentum.

OSHA

In 1968 people outside the safety profession were also asking questions about our lack of progress, notably members of the Congress of the United States. In 1969 a concentrated push was made for federal intervention into the country's occupational safety activities. The push failed, and there was no legislation that year. In 1970, however, it was a different story, for the historic Occupational Safety and Health Act of 1970 did pass and became federal law effective August 28, 1971.

Obviously, many factors are involved in the passage of any new law. No doubt, one reason for the enactment of OSHA was the lack of progress in the safety profession since 1961, a symptom that Congress chose not to ignore. Viewing the symptom of worsening safety records in industry, Congress apparently chose to act quickly, so quickly that no attempt seems to have been made to identify the problem that caused this symptom. Rather than define the problem, Congress jumped to a solution—the passage of a law aimed almost in its entirety at the environment in which people work. Thus our apparent national safety goal for the 1970s, by federal decree, was to improve working conditions. This was apparently based on a congressional conclusion that "things cause accidents." However, this approach was in sharp disagreement with H. W. Heinrich's theory that people, not things, cause accidents, a theory which Heinrich documented with figures from the 1920s and which had been accepted by most professionals.

According to Heinrich's studies, 88 percent of all accidents are caused by people, and only 10 percent are caused by things. While Heinrich's figures are suspect, most professionals have accepted his principle. Accordingly, if OSHA's standards had been perfectly devised (which they were not), if OSHA's administration had been masterful (which it has not been), and if industry had cooperated fully (which it has not), the accident rate in this country could have been reduced by a total of perhaps 10 percent (if we agree with Heinrich's figures).

With less-than-perfect standards, administration, and cooperation, there has been an improvement in the years since OSHA went into effect of about −13 percent (1970 to 1975 frequency-rate comparison). There has been, however, an unprecedented interest in OSHA (and occasionally also in safety) as well as unprecedented expenditures of corporate moneys to promote safety (and lately also of taxpayers' moneys as federal agencies have come under the dictates of OSHA).

We might wonder why, then, with unprecedented interest in OSHA and in safety, there has not been a greater impact on the record. While this question no doubt requires more than a simple answer, it seems that we might well have offsetting results; for each company that makes a gain in safety because of increased interest by management, generated by the passage of OSHA, we see another industry redirecting its corporate time, effort, and money away from safety programming into OSHA compliance.

There now seems to be a distinct trend toward putting OSHA into its proper perspective. Time and effort are again being directed to safety programming. Perhaps we have learned how to deal with OSHA (what to comply with and what to ignore in the standards); perhaps we have learned to ignore it to a certain extent; or perhaps budgets are now adequate to deal with both OSHA compliance and safety programming. I hope the latter is the case.

For many reasons (some touched on above) we shall not spend a lot of time on OSHA in this book, even though compliance with it is important. Every company must comply with OSHA; it is the law.

WHERE WE ARE

This, then, is where we are. We have been involved in industrial safety in an organized fashion since 1911. Originally our emphasis was on improving physical conditions. Since the 1930s we have also considered the unsafe acts of people. Our progress has been excellent—or it was until the 1960s. Since the late 1950s we seem to have slowly lost ground in our battle to control accidents on the job. We are now at a point where we should begin to reexamine our

	1961	*1975*
Frequency rate (number of accidents per million worker-hours)	5.99	13.10
Severity rate (number of days lost per million worker-hours)	666	752
Combined rate	3,989	9,851

(The combined rate, used by some, is merely the product of the frequency and severity rates.)

techniques—and perhaps even our basic beliefs—to ensure that they will carry us to further successes in the future. Let us look at a few fundamental safety tenets.

FUNDAMENTAL SAFETY TENETS

Unsafe Acts and Conditions

From the very beginning we have recognized that certain conditions are involved in accident causation. Thus as soon as it became economically feasible to try to control accidents, we immediately started working on physical conditions. Then in 1931 Heinrich informed us of what now is a painfully obvious and simple truth—that people, not things, cause accidents.

Heinrich stated it this way:

> The occurrence of an injury invariably results from a completed sequence of factors, the last one of these being the injury itself. The accident which caused the injury is in turn invariably caused or permitted directly by the unsafe act of a person and/or a mechanical or physical hazard.[1]

He likened this sequence to a series of five dominoes standing on edge. These dominoes are labeled:

1. Ancestry or social environment
2. Fault of a person
3. Unsafe act or condition
4. Accident
5. Injury

[1]H. W. Heinrich, *Industrial Accident Prevention,* 4th ed., McGraw-Hill Book Company, New York, 1959.

condition, we deal only at the symptomatic level. This act or condition may be the "proximate cause," but invariably it is not the "root cause." To effect permanent improvement, we must deal with root causes of accidents.

The Management System

Root causes often relate to the management system. They may be due to management's policies and procedures, supervision and its effectiveness, training, etc. In our example of the defective ladder, some root causes could be a lack of inspection procedures, a lack of management policy, poor definition of responsibilities (supervisors did not know they were responsible for removing the defective ladder), and a lack of supervisory or employee training.

Root causes are those whose correction would effect permanent results. They are those weaknesses which not only affect the single accident being investigated but also might affect many other future accidents and operational problems.

Accidents and Other Operational Problems

The root causes of accidents (weaknesses in the management system) are also the causes of other operational problems. Though this fact is not immediately obvious, the more we consider it, the more obvious it seems to become. Consider, for instance, how often our safety problems stem from lack of training—and how often our quality problems also stem from this same lack of training. Or consider how poor selection of employees creates safety problems and other management problems. The fundamental root causes of accidents are also fundamental root causes of many other management and operational problems.

This is not a new thought. Heinrich expressed it in a slightly different way:

> Methods of most value in accident prevention are analogous with the methods for the control of quality, cost, and quantity of production.[2]

We believe today that this statement expresses some of Heinrich's best thinking, but for some reason this is the one principle presented by Heinrich that safety people have *not* seemed to live by in the past.

Consider the difference between the way we handle safety and the way we handle quality, cost, and quantity of production. How does management accomplish other things that it wants—a certain level of production, for instance? When management officials decide they want a certain level of production, they first tell somebody what level they want, or they set a policy and definite goals. Then they say to someone, "You do it." They define responsibility. They say, "You have my permission to do whatever is necessary

[2]Ibid.

to get this job done." They grant authority. And finally they say, "I'll measure you to see whether you are doing it." They fix accountability. This is the way management motivates its employees to do what it wants in production, in cost control, in quality control, and in all other areas except safety.

In safety, industry has seemed to take a quite different tack. Management officials have not effectively used the above tools of communication, responsibility, authority, and accountability. Rather, they have chosen committees, safety posters, literature, contests, gimmicks, and a raft of other things that they would not consider in quality, cost, or production. In those other areas management has not worried too much about motivating people. There, management has decided what it wants and then has made sure that it gets exactly that. In safety, we are in the ludicrous position of pleading for management support, instead of advising management on how to better direct the safety effort to attain its specified goals.

The concept that the root causes of accidents are weaknesses in the management system and thus are causes of other operational problems is discussed further throughout this book. It is perhaps best utilized in the tracing system discussed in Chapter 14; this system starts with either an accident or some other management or operational problem and then traces down the root causes.

Severity versus Frequency

Most of us in safety work have always believed in a predictable relationship between the frequency of accidents and their severity. Many studies have been made over the years to determine this relationship, with varying results. Originally, studies seemed to show that for every serious accident we can expect to suffer 29 minor accidents and 300 no-injury accidents or "near misses." This relationship has been discussed and taught for many years. More recent studies show a similar relationship but with completely different numbers.

Common sense dictates totally different relationships in different types of work. For instance, the steel erector would no doubt have a different ratio from that of the office worker. This very difference might lead us to a new conclusion. Perhaps circumstances which produce the severe accident are different from those which produce the minor accident.

Safety professionals for years have been attacking frequency in the belief that severity would be reduced as a by-product. As a result, our frequency rates nationwide have been reduced much more than our severity rates have. One state reported a 33 percent reduction in all accidents between 1965 and 1975, while during the same period the number of permanent partial disability injuries actually increased. This state is typical of others, and its figures are typical of our national figures. In the period 1926 to 1967, the national frequency rate improved 80 percent, while the permanent partial disability rate improved only 63 percent. (This could, of course, be due partly to the changing definition of a partial disability, as laws vary.)

If we study mass data, we can readily see that the types of accidents resulting in temporary total disabilities are different from the types of accidents resulting in permanent partial disabilities or in permanent total disabilities or fatalities. For instance, the National Safety Council (Figure 1-1) shows that handling materials accounts for 25 percent of all temporary total disabilities and for 21 percent of all permanent partial injuries but for only 6 percent of all permanent total injuries and fatalities. Electricity accounts for 13 percent of all permanent totals and fatalities but for a negligible percentage of temporary totals and permanent partials. These percentages would not differ if the causes of frequency and severity were the same. They are not the same. There are different sets of circumstances surrounding severity. Thus if we want to control serious injuries, we should try to predict where they will happen. Today we can often do just that.

The Key Person

Another fundamental tenet of safety states that the supervisor is the key in accident prevention. This seems axiomatic in our thinking. The supervisor is the person between management and the workers who translates management's policy into action. The supervisor has eyeball contact with the workers.

Is this the key person? In a way, yes. However, although the supervisor is the key to safety, management has a firm hold on the keychain. It is only when management takes the key in hand and does something with it that the key becomes useful. Safety professionals have sometimes used the key-person principle to focus their efforts on frontline supervision, forgetting that the supervisor will do what the boss wants, not what the safety specialist preaches.

Management must delegate or assign responsibility for safety down the line, but without losing any of that responsibility itself. This is where many safety programs fall apart: responsibility is assigned to the line organization, but no more thought is given to it by management (until the record turns bad, perhaps). Yet to assign responsibility to the line without fixing accountability is meaningless. This is exactly what we have done for over 40 years in safety, and in many cases we still do it today.

Type of accident	*Temp. total*	*Perm. partial*	*Perm. total*
Handling materials	24.3%	20.9%	5.6%
Falls	18.1	16.2	15.9
Falling objects	10.4	8.4	18.1
Machines	11.9	25.0	9.1
Vehicles	8.5	8.4	23.0
Hand tools	8.1	7.8	1.1
Electricity	3.5	2.5	13.4
Other	15.2	10.8	13.8

FIGURE 1-1 Accident types

THE FUNCTION OF SAFETY

It is only in recent years that most safety professionals have been able to define their role in the safety work that is being accomplished. What they do has changed and will continue to change as our concepts and principles continue to evolve. If permanent results can be effected by dealing with root causes, safety professionals must learn to work well below the symptomatic level.

If accidents are caused by management system weaknesses, safety professionals must learn to locate and define these weaknesses. They must evolve methods for doing this. This may or may not lead them to do the things they did in the past. Inspection may remain one of their tools—or it may not. Investigation may be one of their tools—or it may not. Certainly safety professionals must use new tools and modernize old tools, for their direction is different today—their duties must also be different.

In the safety profession, we started with certain principles that were well explained in Heinrich's early works. We have built a profession around them, and we have succeeded in progressing tremendously with them. And yet in recent years we find that we have come almost to a standstill. Some believe that this is because the principles on which our profession is built no longer offer us a solid foundation. Others believe that they remain solid but that some additions may be needed. Anyone in safety today ought to at least look at that foundation—and question it. Perhaps the principles discussed in Chapter 2 can lead to further improvements in our approach and further reductions in our record.

REFERENCES

Blake, R. P.: *Industrial Safety,* Prentice-Hall, Inc., Englewood Cliffs, N.J., 1943.

Heinrich, H. W.: *Industrial Accident Prevention,* 4th ed., McGraw-Hill Book Company, New York, 1959.

National Safety Council: "Accident Facts," Chicago, 1976.

Petersen, Dan: *The OSHA Compliance Manual,* McGraw-Hill Book Company, New York, 1975.

2

FUNDAMENTALS OF LOSS CONTROL

This chapter discusses some of the more modern theories and principles of a new conceptual framework for industrial safety. In it we shall examine each of these principles briefly to see how it might change our approach to accident prevention in industry.

BASIC PRINCIPLES

Principle 1

An unsafe act, an unsafe condition, and an accident are all symptoms of something wrong in the management system. We know that many factors contribute to any accident. Our thinking, however, has always suggested that we select one of these as the "proximate" cause of the accident or that we select one unsafe act and/or one unsafe condition. Then we remove that condition or act.

The theory of multiple causation suggests, however, that we trace all the contributing factors to determine their underlying causes. For instance, the amputation of a finger in a power press might start with an act (putting the hand under the die) and a condition (an unguarded point of operation). Tracing back from this point might lead, however, into an inquiry concerning why the operator was selected for the job, why the operator was poorly trained, why the supervisor allowed the act, why the supervisor was poorly selected and trained, why the maintenance on the press was poor, why the policy of management allowed an unguarded press, etc.

This principle suggests not that we boil down our findings to a single factor but rather that we widen our findings to include as many factors as seem applicable. Hence, every accident opens a window through which we can observe the system, the procedures, etc. Different accidents would unearth similar things that might be wrong in the same management system. Also, the theory suggests that besides accidents, other kinds of operationsl problems result from the same causes. Production tie-ups, problems in quality control, excessive costs, customer complaints, product failures, etc., have the same causes as accidents. Eliminating the causes of one organizational problem will eliminate the causes of others.

If we were actually to utilize this theory, we would redesign our accident investigation procedures in a way that would enable us to identifiy as many contributing factors as possible in any single incident. Most of the changes would be directed at improving the organizational system—not at finding fault.

Identifying the act, the condition, and the accident only as a starting point to learn why the act and the accident were allowed to happen and why the condition was permitted to exist will lead to effective loss control. We view the accident, the act, and the condition as symptoms of something wrong in the system. Then we try to identify what is wrong in the organizational system that allows an unsafe act to be performed and an unsafe condition to exist.

If in the instance of the finger amputation, we said merely that the cause was the unguarded press, the correction would consist in putting a guard on the press. However, we would have treated only the symptom, not the cause, for tomorrow the press would again be unguarded. This principle applies to any accident. Only when we diagnose causes and treat them do we effect permanent control. The function of the safety professional, then, is similar to that of a physician who diagnoses symptoms to determine causes and then treats those causes or suggests appropriate treatment.

Principle 2

We can predict that certain sets of circumstances will produce severe injuries. These circumstances can be identified and controlled. This principle states that we can predict severity of accidents under certain conditions and thus turn our attention to severity per se instead of merely hoping to reduce it by attacking frequency.

Statistics show that we have been only partially successful in reducing severity by trying to control frequency. National Safety Council figures show an 80 percent reduction in the frequency rate over the last 40 years. The same source shows that during this period there has been only a 72 percent reduction in the severity rate, a 67 percent reduction in the fatal and permanent total rate, and a 63 percent reduction in the permanent partial disability rate.

A number of recent studies suggest that severe injuries are fairly predictable in certain situations. Some of these situations involve:

1. Unusual, nonroutine work. This includes the job that pops up only occasionally and the one-of-a-kind situation. Nonroutine work may arise in both production and nonproduction departments. The normal controls that apply to routine work have little effect in the nonroutine situation.

2. Nonproduction activities. Much of our safety effort has been directed to production work. However, there is a tremendous potential exposure to

loss associated with nonproduction activities such as maintenance and research and development. In these types of activities most work tends to be nonroutine. Since it is nonproduction work, it often does not get much attention from safety, and usually it is not carried out according to standardized procedures. Severity is predictable here.

3. Sources of high energy. High energy sources can usually be associated with severity. Electricity, steam, compressed gases, and flammable liquids are examples.
4. Certain construction situations. Included are high-rise erection, tunneling, working over water, etc. (Actually, construction severity is an amalgam of the previously described high-severity situations.)

These are just a few examples of the areas in which severity is predictable.

Principle 3

Safety should be managed like any other company function. Management should direct the safety effort by setting achievable goals and by planning, organizing, and controlling to achieve them. Perhaps this principle is more important than all the rest. It restates the thought that safety is analogous with quality, cost, and quantity of production. It also goes further and brings the management function into safety (or, rather, safety into the management function). The management function by definition should include safety, but in practice it has not done so. Management has too often shirked its responsibility here. It has not led the way; at best it has given "support."

We in the safety profession are often partly at fault. We have not made management lead the way—only asked for (or hoped for) management support. We have not demonstrated that safety is a management responsibility requiring goal setting, proper planning, good organization, and effective management-oriented controls. At times, we have not even spoken management's language. It is only when management manages safety through its staff safety assistant that we shall see results to reverse our present trends.

Inherent in this principle is the fact that safety is and must be a line function. As management directs the effort by goal setting, planning, organizing, and controlling, it assigns responsibility to line managers and grants them authority to accomplish results. The word "line" here refers not only to first-level supervisors but also to all management-level supervisors above those on the first level, up to the top. Part 2 of this book is devoted to management's role in safety.

Principle 4

The key to effective line safety performance is management procedures that fix accountability. Any line manager will achieve results in those areas in which he or she is being measured by management. The concept of "accountability" is important for this measurement, and the lack of procedures for fixing accountability is safety's greatest failing. We have preached line responsibility for many years. If we had spent this time devising measurements for fixing accountability of line management, we would still be achieving a reduction in our accident record.

A person who is held accountable will accept the given responsibility. In most cases, someone who is not held accountable will not accept responsibility—he or she will devote the most attention to the things that management is measuring: production, quality, cost, or any other area in which management is currently exerting pressure.

This principle is extremely important for the implementation of principle 3. Principle 4, in effect, makes principle 3 work. Several chapters in Part 2 are devoted to accountability. There we discuss its importance in more detail and present some specific techniques for fixing accountability to the line.

Principle 5

The function of safety is to locate and define the operational errors that allow accidents to occur. This function can be carried out in two ways: (1) by asking why accidents happen—searching for their root causes—and (2) by asking whether certain known effective controls are being utilized. The first part of this principle is borrowed from the ideas of W. C. Pope and Thomas J. Cresswell as put forth in their article entitled "Safety Programs Management," in the August 1965 issue of the *Journal of the American Society of Safety Engineers.* This article defines safety's function as locating and defining operational errors involving (1) incomplete decision making, (2) faulty judgments, (3) administrative miscalculations, and (4) just plain poor management practices.

Pope and Cresswell suggest that to accomplish our purposes, we in safety would do well to search out not what is wrong with people but what is wrong with the management system that allows accidents to occur. This thinking is borne out in the ASSE publication entitled "Scope and Functions of the Professional Safety Position," where the position of safety is diagrammed as in Figure 2-1.

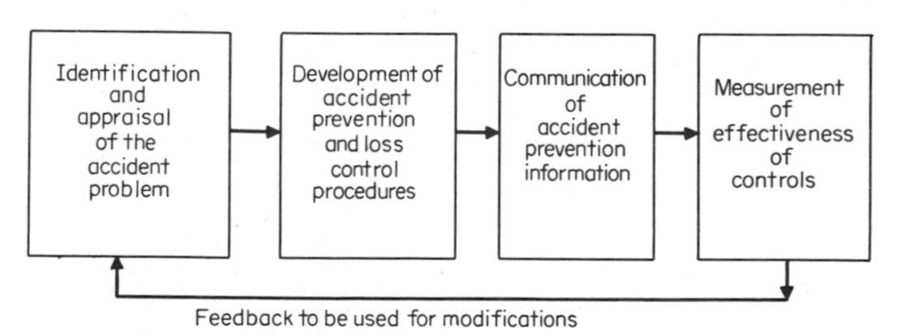

FIGURE 2-1 The professional task

In further describing the four major areas under identification and in appraising the problem, the publication states that it is the function of safety to:

> Review the entire system in detail to define likely modes of failure, including human error, and their effects on the safety of the system. . . .
>
> Identify errors involving incomplete decision making, faulty judgment, administrative miscalculation and poor practices. Designate potential weaknesses found in existing policies, directives, objectives, or practices.[1]

This new concept directs the safety professional to look at the management system, not at acts and conditions.

The second part of principle 5 suggests that a two-pronged attack is open to us: (1) tracing the symptom (the act, the condition, the accident) back to see why it was allowed to occur and (2) looking at the company's system (procedures) and asking whether certain things are being done in a predetermined manner that is known to be successful.

We have discussed item 1 above: tracing the symptom back to its underlying root causes. The accident described in Chapter 1 illustrated the difference between removing the symptom (the act and the condition) and asking why the act was allowed to be performed and the condition was permitted to exist (tracing back to root causes).

We have not yet discussed item 2 above: looking at the system to determine what controls are in effect or asking whether the company has the controls that are needed for the exposures it faces. Much of the rest of this book pertains to this second approach to safety: looking at the system. The content of the chapters suggests what to ask "whether" about. It suggests that you analyze your company by asking questions in the areas of:

Management's safety policy

How your company is organized

What the function of your safety department is

How safety is niched into the organization

What the relationships between staff people are

How line safety responsibilities are defined

How management holds people accountable

How supervisors are measured in safety

How employees are selected

How employees are trained

[1]American Society of Safety Engineers, "Scope and Functions of the Professional Safety Position," Chicago, 1966.

How supervisors are motivated

How management is motivated

In discussing some of these topics, I shall attempt to provide readers of the book with outlines for thinking so that they may take an objective look at their own companies—and analyze their own needs.

THE ROLE OF SAFETY

Before we begin our discussion of managing the safety function and motivating for safety, let us define once more the role of safety—where it fits in an industrial enterprise.

For many years the slogan of the National Safety Council was "safety first," and many professionals believed and preached this. Today we realize that we really do not want "safety first" any more than we want "safety last." In other words, we do not want to think of safety as being separate from the other aspects of production. Obviously, we want effective production first, but we want it to be accomplished in such a manner that no one is hurt and losses are minimized.

In the past, safety professionals were oriented to safety programs for their companies—the aim was to superimpose a safety program on the organization. Today, safety professionals realize that what is really needed is "built-in" safety, "integrated" safety, not an artificially introduced program. Safety must be an integral part of a company's procedures. That is:

> We do not want production and a safety *program,* or production *and* safety, or production *with* safety—but, rather, we want *safe production.*

We can better understand the concept of integrated safety if we discuss how safety fits into production. The goal of management is efficient production—production which maximizes profit. To obtain this goal, it has two areas of basic resource: (1) employees and (2) facilities, equipment, and materials (Figure 2-2). Management brings many influences to bear upon both these resources (Figure 2-3). To the company's personnel, management applies such influences as training, selection and placement processes, employee health programs, and employee relations practices. The resource of facilities, equipment, and materials is influenced by maintenance, research, and engineering.

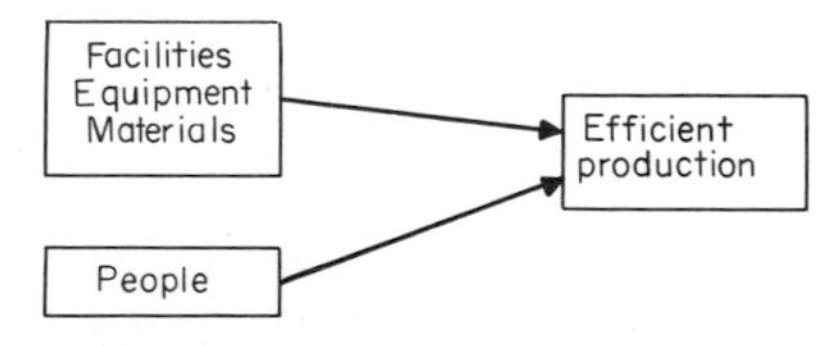

FIGURE 2-2 The role of safety

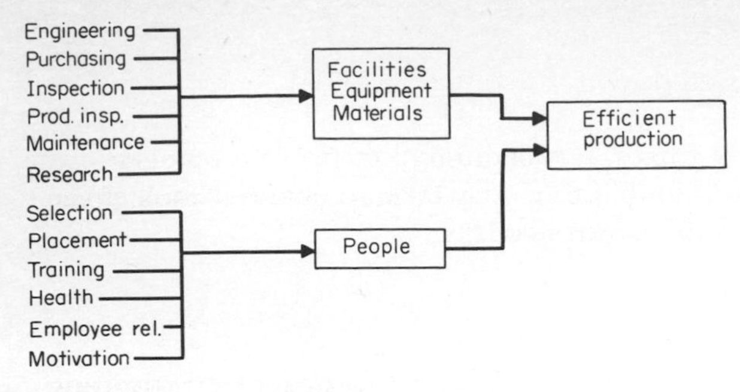

FIGURE 2-3 The role of safety

These influences and basic resources are brought together through various procedures (Figure 2-4). The function of safety is to:

1. Build safety into these procedures
2. Continually audit the carrying out of these procedures to ensure that the controls are adequate

Safety accomplishes these tasks by asking *why* certain acts and conditions are allowed and asking *whether* certain known controls exist.

Safety is not a resource; it is not an influence; it is not a procedure; and it certainly is not a "program." Rather, safety is a state of mind, an atmosphere that must become an integral part of each and every procedure that the company has. This, then, is what we mean by "built-in" or "integrated" safety. It is the only brand of safety that is permanently effective.

Since any accident, unsafe act, or unsafe condition indicates a system failure, the safety professional must become a systems evaluator. The following chapters are intended to be an aid in this task.

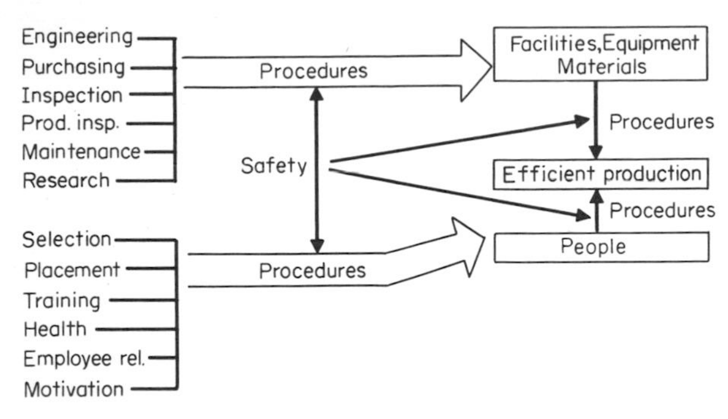

FIGURE 2-4 The role of safety

REFERENCES

American Society of Safety Engineers: "Scope and Functions of the Professional Safety Position," Chicago, 1966.

National Safety Council: "Accident Facts," Chicago, 1976.

Pope, W. C., and **T. J. Cresswell:** "Safety Programs Management," *Journal of the ASSE,* August 1965.

PART TWO

MANAGING SAFETY PERFORMANCE

In Chapter 2 it was suggested that one way to achieve good safety performance is to observe our industrial organizations and determine whether they are set up to do certain things—things we have found to be successful.

Part 2 will acquaint the reader with the questions that should be asked about the organization. We shall look chiefly at these broad management areas to see how they influence accident prevention:

Policy—responsibility and authority

Accountability

Measurement and appraisal

3

MANAGEMENT'S DIRECTION OF SAFETY

Our thinking on "managing the safety function" must start with a written safety policy—a definition, if you will, of management's desires concerning safety. Most executives agree that a policy is a fine thing to have, but few agree on what a policy is. This is borne out by the flood of manuals and handbooks in which "policy" is indiscriminately interchanged with "practices," "procedures," and even "rules."

"Policy" is often confused with "rules," "established practices," "procedures," and "precedents"—not only in speech but also in action. But "policy" has certain unique implications. It implies scope for discretion, initiative, and judgment in deciding what ought to be done in specific situations.

Too often in discussions regarding policy, we become bogged down trying to sort out policy from procedures, from SOPs and from rules. Perhaps it is not all that necessary for policy to be pure. The most important thing is probably whether management's interest is accurately communicated.

DEFINITION OF POLICY

One dictionary says that policy is "a settled course adopted and followed by a body." When top executives determine and announce a "settled course," they are affirming a shared purpose or one in which they want to enlist the voluntary cooperation of every member of the organization. They are giving out a guide for thinking, and this guide is to be used by other members who have been delegated authority to make decisions in keeping the company on a given course while realizing company purposes.

A policy should do three things:

1. It should affirm long-range purpose.
2. It should commit management at all levels to reaffirm and reinforce this purpose in daily decisions.

3. It should indicate the scope left for discretion and decision by lower-level management.

WHY SAFETY POLICY?

Up to this point our discussion of policy has pertained to all kinds of management directives. What about safety policy—what is distinctive about it? I believe that written safety policy is important in any organization—that it is the starting point for all our various activities in a safety program.

If the principle *"Safety is a line responsibility"* is true, it is important not only that we as safety people believe it but also that the line organization believe it. The line organization will believe that safety is its responsibility *only* when safety is definitely assigned to it by management.

Another principle is that *"Management should direct the safety effort by setting safety goals and by planning, organizing, and controlling to achieve them."* A safety policy is management's expression of direction to be followed. It is management's first step in organizing to accomplish its desires. In almost all cases it is important that management's safety policy be in writing to ensure that there will be no confusion concerning direction and assignment of responsibility.

Why do I urge that safety policy be put in writing when other company policy is not? Why do I propose that safety policy be publicized for all employees when other policies are merely put in the policy manual? For this reason: Safety policy, more than most other policies that come from management, requires some action from each individual in the organization, from the president to the lowest-rated worker. Safe performance of an organization requires that a decision be made by each person in it. The most important factor that each will consider in making the decision for or against safety is, "What does the big boss want from me?"

It is only through policy (written and well publicized) that the worker on the line or on the construction job will know management's desires concerning safety. This is why, regardless of how most company policy is handled, safety policy must be put in writing and publicized at least to the point where each and every employee is aware of it. True, it should be part of the organization's operations manual—but it should also be well known to each person in that organization, and this usually requires some special emphasis.

Many success stories could be related that start with a statement of management safety policy. The policy used in one success story is included in Appendix B. The company involved had a safety record that was, at best, about average. A new safety director was appointed. He was inexperienced but anxious to try, and he started right in on his new job. His first effort was to crystallize his thoughts into a safety policy. He submitted this to the manager in charge of his location, and, much to his surprise, the manager liked what he read. The manager put his name to the policy and published it. From that moment the company achieved an enviable record of over 2½ million worker-hours without a disabling injury.

In the first year under this policy this location received a substantial refund from its insurance carrier for its excellent record—more than enough to offset any expenditures in safety. Obviously, this record took more than a policy alone. However, the policy set the stage for everything else that was done, and the record could not have been achieved without it.

WHAT SHOULD BE INCLUDED?

What is included in the safety policy may vary from company to company. No doubt most organizations will not write "pure" policy. They will include, either intentionally or inadvertently, some procedures, some philosophy, and perhaps even some rules, with their expression of management will. This is perfectly all right—whatever serves the company best is what should be included. Appendix B presents several different safety policies, reprinted with the permission of the authors. The reader will note that they differ markedly in content and vary considerably in style.

No one policy is right or wrong—we might best assume that each is right for the organization it serves. We can, however, outline some of the things that should be included in most management policies on safety. As a minimum, the following areas ought to be touched on in a safety policy:

1. Management's intent. What does management want?
2. The scope of activities covered. Does the policy pertain only to on-the-job safety? Does it cover off-the-job safety also? Fleet safety? Public safety? Property damage? Fire? Product safety?
3. Responsibilities. Who is to be responsible for what?
4. Accountability. Where and how is it fixed?
5. Staff safety assistance. If there is staff safety, how does it fit into the organization? What should it do?
6. Safety committees. Will there be committees? What will they do? Why do they exist?
7. Authority. Who has it, and how much?
8. Standards. What rules will the company abide by?

These questions may provide some insight to the reader who is contemplating writing a safety policy. There is one more procedure to be followed: To be effective, the policy must be signed by management. It may be conceived and written by the safety professional, but it *must* be published under the name of the executive who is responsible at least for all other production activities. It is often important that the *top* executive issue safety policy, as safety affects *all* departments in the organization—not just production.

Responsibility, Authority, and Accountability

The three terms "responsibility," "authority," and "accountability" are closely related. They are, in fact, sometimes used interchangeably, although they should not be. For our purposes in this book they are defined as follows:

RESPONSIBILITY. The fact of having to answer for activities and results in safety.

AUTHORITY. The right to correct, command, and determine courses of action.

ACCOUNTABILITY. The fact of active measurement by management to ensure compliance with its will. In defining accountability for accident prevention, we speak of management doing something to ensure action.

It has often been said that the three must go together: that a person who is given responsibility and is held accountable should also be given authority. This is an oversimplification, however, and is seldom followed in practice.

ORGANIZATION

One of the functions of the safety policy is to set up the safety organization for the business. Organization has already been touched on in the discussion of the basic principle, *"Safety is a line function,"* which actually concerns setup.

Before going further, we should make sure that we have a clear understanding of the difference between line and staff functions:

LINE FUNCTIONS. Doing the work for which the enterprise exists and making the operational decisions (issuing orders, etc.) that will get the work done

STAFF FUNCTIONS. Assisting, supporting, planning, and facilitating the work of the line, but without authority to command or direct the line

The Staff Safety Specialist

Obviously, since the line has primary safety responsibility, the safety specialist (or whatever title is used) is and must be staff. As staff, the safety specialist has no responsibility for the safety record or the results. The responsibilities of the job involve activities which *help* the line achieve its goals in safety. The safety specialist has no authority over the line. He or she may have a great deal of influence, but this is quite different from authority. How much influence or power the safety specialist has will depend on the organization and on the personality of the individual in the line position. The safety professional in any organization obtains results by using either of two methods: (1) making a recommendation to a line executive who issues an order, or (2) obtaining acceptance of his suggestions voluntarily from line supervisors without taking the chain-of-command route. More often than not, the purposes of the safety specialist are accomplished by the second route, and the first route is used only for emergencies.

Most line managers realize that the safety specialist does have stronger influence than that shown on the organization chart. This staff specialist is an expert in the field, has certain status, has management's attention and backing, and, if worse comes to worst, has some influence on management's appraisal of the line manager and hence on this person's future. So, although the safety specialist has no authority, the position is not without power. There are situations also where the safety specialist is given degrees of authority over the line.

It may be stated in the policy that, in certain situations, the safety specialist must be consulted, must give his or her approval, or can even temporarily step in and assume command (stop the operation). This granting of temporary authority to the staff safety specialist in industry is common and right. Even so, we must keep in mind that basically we work through influencing the line, not by directing it. Any time that safety must step in and assume direction of the line, this indicates a failure in that the line has not taken care of the situation first.

Where to Install Safety

To whom should the staff safety specialist report? Should the specialist be staff to a line manager or staff to a staff executive? These questions have been debated for years, and we are sure of only one thing—that there is no one right answer. It depends on the organization and the personalities of the people involved. There are, however, some criteria that can be used in assessing the right place for safety in an organization.

Any discussion of organization charts must be qualified by the observation that charts do not reflect the give-and-take of powerful executives. Whether safety should report to line or to staff might depend on a more rational process of structuring the organization to achieve goals, or it might depend only on personalities.

The location of safety within the organization must reflect the fact that root causes of accidents exist at all levels, in all departments, and in all functions. Interrelated causes can be in maintenance, purchasing, tool-crib control, selection of personnel, etc. Therefore, safety seeks to exert a control, or at least an influence, on every department head, every function, and every supervisor.

We can begin to ask some questions about company A in (Figure 3-1). Presumably, if the works manager demands accident control from superintendents, the safety director could be effective at least in the area controlled by the superintendents. But can the safety director exert effective influence on people who are wrecking company vehicles, or on people in the R & D laboratory? If the works manager is apathetic about safety, could the safety director even

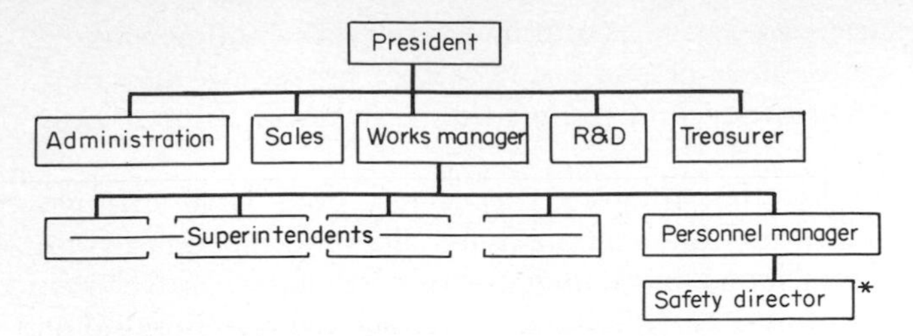

FIGURE 3-1 Organization chart, company A

reach the superintendents? How would it be possible to establish liaison and understanding with the treasurer, who buys the insurance and who best realizes the dollar cost of accidents? Note that personnel is a staff function to the works manager, not to the whole organization. Could personnel exert standards of driving competence in the hiring of salespeople? What, if any, difference would it make if safety were assigned to report directly to the line works manager instead of the staff personnel manager?

In company B (Figure 3-2) safety reports to a staff function, industrial relations, which is staff to the whole organization, not to just one line department. Presumably, the safety director, through his or her executive boss, now has a channel for communicating with the R & D department (or the sales department, purchasing, maintenance, etc.). In any case, the boss is close to the ear of the chief executive. This proves nothing yet about where safety should be installed, but we have arrived at a principle: *The voice of the safety director is the voice of the boss.* Who that boss is, how far the boss's voice reaches, and whether the boss speaks for safety are among the factors that limit the safety director's effectiveness. Is the boss in line or staff? If staff, to whom is the boss staff? These are factors to contend with. We cannot say that safety must of necessity report high or low, staff or line. But we can now present criteria for determining how safety is allocated in the organization:

1. *Report to a boss with influence.* In part this is a personal evaluation; in part it is an evaluation of the structure of the organization. If the boss is line, does his or her line authority encompass the hazards to be controlled? If

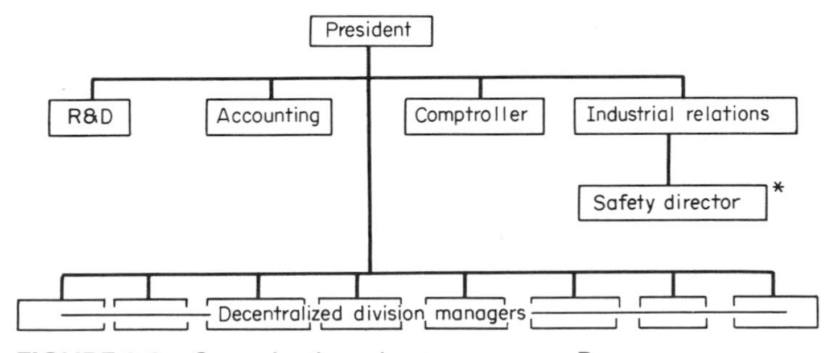

FIGURE 3-2 Organization chart, company B

staff, can the boss's voice reach an executive whose command will buttress necessary action?

2. *Report to a boss who wants safety.* This follows inevitably. Problems arise when a chief executive wants results but the voice of the safety director is muffled by an immediate boss who is concerned with other problems.

3. *Have a channel to the top.* Management properly sets the priorities between production results and safety results, between sales expansion and elimination of unsafe driver-salespeople, between security of confidential research and the prying eyes of the safety professional. This is not to say that safety must be placed in the upper echelons, but it does assert that all parts of the organization must have a channel to the upper echelons. Too often the only channel is a bypass, with all its frictions.

4. *Perhaps, install safety under the executive in charge of the major activity.* The safety function in this case serves as staff to that executive. This obviously eliminates the "channel to the top." Nonetheless, if the acute need is in the shop, let the safety specialist work with the shop executive on their control. If the acute need involves truck operation, let the transportation executive handle the problem with the safety director as staff assistant. Influence in other departments may be weak, but management has directed its control to the spot that hurts.

These criteria serve more to assess present structure than to determine where safety should be located. The latter depends on management goals. This is illustrated in Figures 3-3 and 3-4. In company C (Figure 3-3) note the allocations for fire, security (guards), and health. This arrangement makes sense when we realize that this company manufactures a nontoxic but flammable household product, but has recently gone into insecticides and other products of high toxicity.

Fire control is concentrated where it is needed, rather than as a part of either safety or security. Other functions of security (theft, vandalism, sabotage,

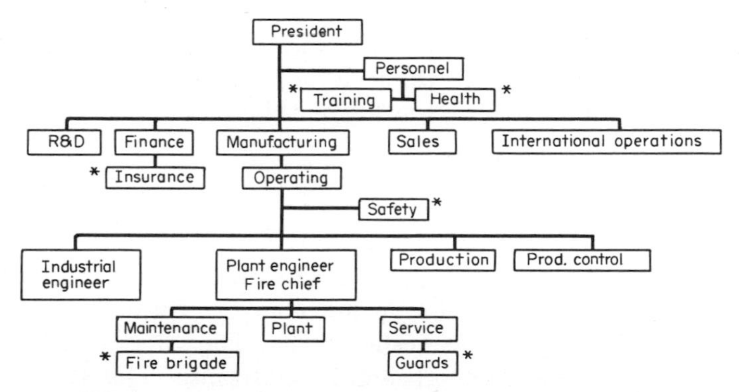

FIGURE 3-3 Organization chart, company C

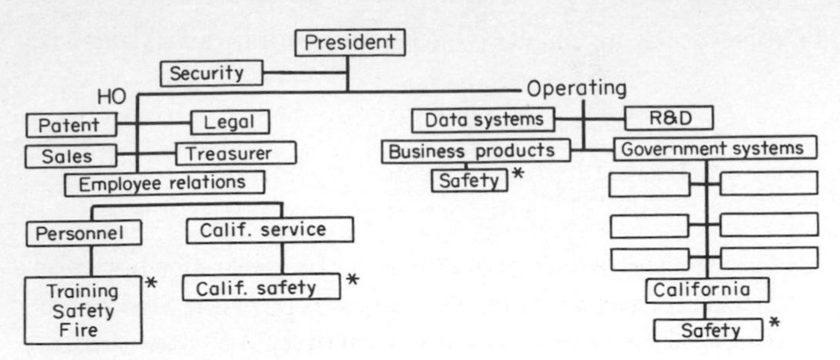

FIGURE 3-4 Organization chart, company D

confidential information) demand little management attention, and the security function becomes simply one of "guarding." Safety is staff to a line executive. Note "health," however. Here management must know what problems will arise with any new product. Health and control of health hazards are situated to serve top-echelon decision making. In short, the chart reflects management control of its problems and a sense of management goals.

This principle is further revealed in company D (Figure 3-4). Here, the left branch portrays an East Coast home office. The right branch portrays a number of locations across the United States, including a California operation. It sells "brainpower," primarily on government contracts, but it also has a manufacturing operation called "business products." The chart, of course, does not reveal the reason for the multiple niches of safety, but it reveals something. The California operation needs safety on the spot, but 3,000 miles away it also needs safety as part of "California service" of the home office. The niche for security is understandable when we realize that a breach of security could jeopardize the whole existence of the company.

Organization structure begins with management objectives and goals. The tug and pull of contending executives change and distort the structure, but even this reflects management goals or accommodation of goals. Though business affairs cannot be constrained or described in neat little boxes, no organization is totally unstructured.

Structure reflects grouping of activities, but the logic of grouping is the logic of goals, not the logic of words. Safety has been effectively allocated into line or into staff, grouped with fire or security, or the legal department, or maintenance, or personnel, or even the training department. The practice of other organizations is a poor guide to follow.

The Part-Time Specialist

What has been said in this chapter refers not only to the full-time safety professional but also to the personnel specialist, the line manager, the insurance manager, and others with a part-time safety responsibility. The safety professional is still staff and works through influencing, not directing. The safety professional's niche in the organization still affects his or her effectiveness.

There are, of course, far more part-time safety specialists in American industry today than full-time safety professionals. It would seem that we should be able to speak of special rules that apply to the part timer: duties associated with the position. Actually, however, this is difficult because basically the role should be no different from that of the full-time safety professional. The function is still that of advising and assisting the line organization in its job of loss control. Hence, anything that has to do with loss control concerns the part-time specialist as much as it does the full-time professional.

The problem is that when safety is a part-time function, it is too often a "tacked-on" function; it is the secondary job, never the primary one. Perhaps this is a necessary evil, but to admit that is to admit that safety is less important than other functions. In most cases, management's policy will say otherwise. Perhaps here again the part-time safety specialist should share the fault for not holding management to its own stated policy when the time squeeze is on.

The outside consultant can give valuable assistance to the part-time safety specialist. Too often the consultant wants to help but is not used properly by the part-time safety specialist. Sit down with this person, whether he or she is from the government or the insurance field or is a private consultant, and develop an action plan. If used properly in the areas of their specialty, consultants can end up as staff assistants to part-time specialists. This often becomes the equivalent of a full-time function. Consultants often specialize in those areas in which the part-time specialist needs help the most: supervisory training, industrial hygiene, cost analyses, etc.

Often the primary reason why the part-time specialist is so much less effective than the full-time safety professional is simply lack of knowledge. This is hardly excusable, for ample information on safety is available.

The Safety Committee

The industrial safety movement has been blessed with (and cursed by) an organizational phenomenon that has affected no other segment of industry. This is the safety committee. The marriage of industrial safety and the committee approach is, to this day, so strong that in some states companies are literally forced into forming safety committees through codes or through insurance rating plans. Government publications urge committees.

How and why committees were initially used in safety is a mystery. They have not been used to the same extent in any other management function. Many organizations utilize committees in their executive ranks, but safety has few executive committees and many supervisor and worker committees.

Although many competent professionals use committees, some extensively, no competent professional depends on committees to do the safety job alone. At best, committees provide training and motivation for their members. At worst, they are a total waste of time.

Whether or not an organization wishes to use safety committees is strictly its own decision, and to say that committees are essential to safety results is ludicrous. They may help the safety director; they may not. This will depend on many things.

If management takes care of the basic management functions of preparing policy, fixing accountability, training supervisors and employees, selecting and motivating employees well, etc., then safety committees probably are not necessary.

On the other hand, if management *does not* effectively carry out these functions, then using safety committees certainly will not ensure results. In this kind of situation it is doubtful whether having committees will help at all.

Although safety committees are extremely common in industry today, it is difficult to imagine a situation to which there would be a better solution than the more normal, more effective one of good management-directed control. If safety committees are used, the safety professional would do well to examine closely the workings of those committees. For effective operation these rules are essential:

1. Define the duties and responsibilities clearly.
2. Choose members in view of these duties.
3. Provide any necessary staff assistance.
4. Design procedures for prompt action.
5. Choose the chairperson carefully.

The Safety Specialist's Job

We have not yet discussed the actual duties and responsibilities of the staff safety specialist, whether working full or part time. Actually, we cannot. Safety specialists perform duties that they have set for themselves. The job of staff safety specialist is very much a self-defined one. The duties will vary, depending on the size of the organization, the number of locations worked with, the operations themselves, the people above the safety specialist and in line management, the problems presently facing the company, the other staff people and specialists available, and where the safety specialist fits into the organization.

In 1963 the executive committee of the American Society of Safety Engineers initiated a plan designed to identify the type of work that safety personnel should be doing. This was part 1 of a three-part major project; it described the scope and functions of the safety professional's position. Part 2 developed a curriculum or formal course of study, leading to a university degree, which would prepare the safety professional to perform the functions described in part 1. Part 3 established procedures leading to the acquisition of some form of certification or registration as a means of demonstrating competence in the field. Following are the results of part 1 of the project.[1]

The major functions of the safety professional are contained within four basic areas. However, application of all or some of the functions listed below will depend upon the nature and scope of the existing accident problems, and the type of activity with which he is concerned.

The major areas are:

A. Identification and appraisal of accident- and loss-producing conditions and practices, and evaluation of the severity of the accident problem

B. Development of accident prevention and loss-control methods, procedures, and programs

C. Communication of accident and loss-control information to those directly involved

D. Measurement and evaluation of the effectiveness of the accident and loss-control system and the modifications needed to achieve optimum results

A. IDENTIFICATION AND APPRAISAL OF ACCIDENT AND LOSS-PRODUCING CONDITIONS AND PRACTICES, AND EVALUATION OF THE SEVERITY OF THE ACCIDENT PROBLEM

These functions involve:

1. The development of methods of identifying hazards and evaluating the loss-producing potential of a given system, operation or process by:

   a. Advanced detailed studies of hazards of planned and proposed facilities, operations, and products

   b. Hazard analysis of existing facilities, operations, and products

2. The preparation and interpretation of analyses of the total economic loss resulting from the accident and losses under consideration

3. The review of the entire system in detail to define likely modes of failure including human error and their effects on the safety of the system:

   a. The identification of errors involving incomplete decision making, faulty judgment, administrative miscalculation, and poor practices

   b. The designation of potential weaknesses found in existing policies, directives, objectives, or practices

4. The review of reports of injuries, property damage, occupational diseases, or public liability accidents, and the compilation, analysis, and interpretation of relevant causative-factor information.

   a. The establishment of a classification system that will make it possible to identify significant causative factors and determine trends

[1]American Society of Safety Engineers, "Scope and Functions of the Professional Safety Position," Chicago, 1966.

   b. The establishment of a system to ensure the completeness and validity of the reported information

   c. The conduct of thorough investigation of those accidents where specialized knowledge and skill are required

5. The provision of advice and counsel concerning compliance with applicable laws, codes, regulations, and standards

6. The conduct of research studies of technical safety problems

7. The determination of the need of surveys and appraisals by related specialists such as medical, health physicists, industrial hygienists, fire protection engineers, and psychologists, to identify conditions affecting the health and safety of individuals

8. The systematic study of the various elements of the environment to assure that tasks and exposures of the individual are within his psychological and physiological limitations and capacities

B. DEVELOPMENT OF ACCIDENT PREVENTION AND LOSS-CONTROL METHODS, PROCEDURES, AND PROGRAMS

In carrying out this function, the safety professional:

1. Uses his specialized knowledge of accident causation and control to prescribe an integrated accident and loss-control system designed to:

   a. Eliminate causative factors associated with the accident problem, preferably before an accident occurs.

   b. Where it is not possible to eliminate the hazard, devise mechanisms to reduce the degree of hazard.

   c. Reduce the severity of the results of an accident by prescribing specialized equipment designed to reduce the severity of an injury should an accident occur.

2. Establishes methods to demonstrate the relationship of safety performance to the primary function of the entire operation or any of its components.

3. Develops policies, codes, safety standards, and procedures that become part of the operational policies of the organization.

4. Incorporates essential safety and health requirements in all purchasing and contracting specifications.

5. As a professional safety consultant for personnel engaged in planning, design, development, and installation of various parts of the system, advises and consults on the necessary modification to ensure consideration of all potential hazards.

6. Coordinates the results of job analysis to assist in proper selection and placement of personnel whose capabilities and/or limitations are suited to the operation involved.

7. Consults concerning product safety, including the intended and potential uses of the product as well as its material and construction, through the establishment of general requirements for the application of safety principles throughout planning, design, development, fabrication, and test of various products, to achieve maximum product safety.

8. Systematically reviews technological developments and equipment to keep up-to-date on the devices and techniques designed to eliminate or minimize hazards, and determines whether these developments and techniques have any applications to the activities with which he is concerned.

C. COMMUNICATION OF ACCIDENT AND LOSS-CONTROL INFORMATION TO THOSE DIRECTLY INVOLVED

In carrying out this function the safety professional:

1. Compiles, analyzes, and interprets accident statistical data, and prepares reports designed to communicate this information to those personnel concerned.

2. Communicates recommended controls, procedures, or programs, designed to eliminate or minimize hazard potential, to the appropriate person or persons.

3. Through appropriate communication media, persuades those who have ultimate decision-making responsibilities to adopt and utilize those controls which the preponderance of evidence indicates are best suited to achieve the desired results.

4. Directs or assists in the development of specialized education and training materials and in the conduct of specialized training programs for those who have operational responsibility.

5. Provides advice and counsel on the type and channels of communications to ensure the timely and efficient transmission of usable accident prevention information to those concerned.

D. MEASUREMENT AND EVALUATION OF THE EFFECTIVENESS OF THE ACCIDENT AND LOSS-CONTROL SYSTEM AND THE NEEDED MODIFICATIONS TO ACHIEVE OPTIMUM RESULTS

The safety professional:

1. Establishes measurement techniques such as cost statistics, work sampling, or other appropriate means, for obtaining periodic and systematic evaluation of the effectiveness of the control system.

2. Develops methods that will evaluate the costs of the control system in terms of the effectiveness of each part of the system and its contribution to accident and loss reduction.

3. Provides feedback information concerning the effectiveness of the control measures to those with ultimate responsibility, with the recommended adjustments or changes as indicated by the analyses.

On the basis of this outline and his or her own definition of the needs of the organization, the safety professional must define the duties and responsibilities of the job.

OSHA Responsibilities

After defining the safety responsibilities of the job, the safety director must go back and define the OSHA responsibilities. This process is not often similar to that of defining safety tasks. The safety director's role in OSHA compliance is occassionally self-defined, but usually it is not. In terms of OSHA compliance, the safety director assumes a quite different role, that of the corporate OSHA expert—the only one who 'knows" the standards. Probably the reason for this is that the standards are written in such a way that the line organization perceives them as being much more technical and difficult to understand than they really are. As a result, the safety specialist is not able to ask line managers to ensure that their departments are in compliance.

Obviously, it did not have to be this way. There is no reason why a line manager could not ensure compliance if the relatively simple concepts covered by the standards had been simply and understandably written.In any event, the safety director seems to be stuck with a role in OSHA compliance that can usually be described as follows:

1. The safety specialist must interpret the physical standards for all in the company.

   Most line managers are simply not willing to expend the time and energy necessary to interpret the 250-odd pages of "federalese."

2. The safety specialist must sift through those pages to find the various administrative requirements tucked in there (there were some 367 such requirements at last count). Personal experience tells me that no line manager would do this.

3. The safety specialist must then find the physical and administrative violations existing in the organization.

4. The safety specialist must then assign priorities to the violations found and begin to schedule corrections.

5. The safety specialist must ensure that the corrections are accomplished.

6. The safety specialist must then construct a plan to assure that the organization will stay in compliance with the law.

7. Finally, the safety specialist must document everything that has been done to achieve compliance (and perhaps some things that have not been done). Documentation avoids fines and also shows good faith, resulting in the reduction of fines. It also helps ease the aftershock that sometimes

characterizes a closing conference between the OSHA compliance officer and the corporate officers.

REFERENCES

American Society of Safety Engineers: "Scope and Functions of the Professional Safety Position," Chicago, 1966.

Heinrich, H. W.: *Industrial Accident Prevention,* 4th ed., McGraw-Hill Book Company, New York, 1959.

"**Organization:** Simplified Definitions of Staff," Northwestern University, The Traffic Institute, Evanston, Ill., 1962.

Petersen, Dan: *The OSHA Compliance Manual,* McGraw-Hill Book Company, New York, 1975.

Pigors, Paul and **Faith Pigors:** "Let's Talk Policy," *Personnel,* July 1950

Strode, Marvin: "How to Motivate Employees to Work 2,000,000 Safe Man Hours," presented before the Pulp and Paper Section of the 1965 National Safety Congress.

Weaver, D. A. and **D. C. Petersen:** "Criteria to Niche Safety," *Industrial Security,* August 1966.

4

ACCOUNTABILITY

Chapter 2 asserted (principle 4) that *"The key to effective line safety performance is management procedures that fix accountability."* It was also stated that this principle is the most important one in safety (and the one most often overlooked).

Most safety professionals today believe in the importance of written management policy concerning loss control. We realize the need for management to express its will in this area more than in any other. Without such written guidelines, signed by the top executive, there is little hope for line compliance with safety programs. Safety affects the entire organization; hence there is a need for a definite policy which all managers can use to control their problems.

The following are generally accepted as being the purposes of a written policy:

1. To assert management's will
2. To state responsibilities and authorities
3. To specify what staff assistance will be provided

The average company, even when it writes a safety policy, does not design any procedures to fix accountability, that is, to measure the safety effectiveness of line managers' performance. Until such procedures are designed, safety cannot be accomplished. Furthermore, we often stop with the writing of the policy, which is really just a beginning point. A written policy is just a piece of paper and is meaningless by iself; only when a company succeeds in getting its people to live by the policy does it become effective.

It has been estimated that fewer than 10 percent of the companies in the United States have written safety policies. A study of 74 safety policies of companies in all types of business throughout the United States revealed that only one even mentioned the idea of fixing accountability to the line organization. On the basis of these figures, only 13 companies in every 10,000 actually do fix accountability for safety to the line.

For well over 50 years we have been preaching the principle of line responsibility in safety work, and yet there are still supervisors today who say, "Safety is the safety director's job" or "If that's a safety problem, take it up with the safety committee." Worse yet, in many companies, when an accident occurs, it goes on the safety specialist's record instead of on the record of the line supervisor in the department where it occurred. Thus, instead of simply preaching that the line has responsibility, we should have been devising procedures to *fix* such accountability. When a person is held accountable (is measured) by the boss for something, he or she will accept the responsibility for it. If not held accountable, he or she will not accept the responsibility. Effort will be put forth in the area in which the boss is measuring.

In safety work, there are three ways of fixing responsibility. We can fix accountability on the basis of the line's activities, we can fix accountability on the basis of the results of those activities, or we can use both methods.

ACCOUNTABILITY FOR RESULTS

Measurement Tools

Charging Accidents to the Department Figure 4-1 shows a partial listing of the things that we might consider measuring in fixing accountability for results. One of the simplest means of doing this is simply to charge accidents to the department in which they occurred. If a person in supervisor A's department suffers a disabling injury, it shows up on supervisor A's record. An adaptation of this is to charge the claim costs back to the line. Here we are measuring the line supervisor in terms of *dollars,* which is a far better measuring stick than any other that we have today. The dollar sign means something to that supervisor and to management. Every accident can go directly into the profit and loss statement of the supervisor, who then will pay for the accidents; the cost will come out of the supervisor's working budget.

Prorating Insurance Premiums Some companies choose to prorate their insurance premiums. When an accident occurs, the insurance company pays the direct costs of that accident. However, the insurance premium that the company pays is influenced by what the costs of the accidents are. If a depart-

ACCOUNTABILITY FOR RESULTS

(1) Charge accidents to departments

(a) Charging claim costs to the line

(b) Including accident costs in the profit and loss statements

(2) Prorate insurance premiums

(3) Put safety into the supervisor's appraisal

(4) Have safety affect the supervisor's income

FIGURE 4-1

ment or a location is charged a specified amount of the insurance premium based on its percentage of the total accident costs, this will be a more accurate dollar measuring stick.

Let us assume, for example, that the XYZ Manufacturing Company has three plants; plant A has 2,000 employees, plant B has 500 employees, and plant C has 5,000 employees. Prorating the insurance premium on the basis of the way the plant used it suggests that, regardless of size, they be charged the proportion of that premium on the same rate as their losses. The XYZ Manufacturing Company insurance premium is $100,000. Thus:

	Accident loss record	*Premium charged*
Plant A	$15,000, or 30%	$30,000
Plant B	$20,000, or 40%	$40,000
Plant C	$15,000, or 30%	$30,000

It will be difficult for plant B to show a profit this year because of its large accident losses.

Safety Included in the Appraisal When line supervisors are appraised on safety as well as on their production records, they generally become far more interested in accident prevention and begin to do something about it. This factor is too often overlooked today in the appraisal of line managers.

Losses Affect Pay When line supervisors' paychecks are in some way influenced by their accident records as well as by their production records, they begin to think of safety as a prime consideration. This has been used very effectively in a number of construction companies. When a job superintendent must subtract accident costs from the profit figure for a job and when the superintent's personal bonus for the job is reduced or wiped out because of this, that superintendent rapidly becomes more interested in accident prevention in the future.

All the above are keyed to the idea that the accident record is the line supervisor's—not the safety department's or the insurance manager's.

The Dollar

Two of the above techniques utilize the dollar as the measuring stick, instead of measuring supervisors in terms of number of accidents, number of days lost, or the commonly used frequency rate or severity rate. Many people today believe that the dollar is a far better measuring stick than any other in safety, and many companies are beginning to utilize it effectively. In the past, it was difficult if not impossible to utilize the dollar for the following reasons:

1. In the case of serious accidents, actual claim costs are not available for a long period of time—in some cases, years after the accident happened.

2. There is no way to convert a frequency rate or a severity rate into dollars.
3. If a company operates in several states, the actual claim costs will be an unfair measuring stick because the benefits vary so much from state to state.

Today, it is possible to fix dollar accountability by using a system in which the costs are estimated by a predetermined formula based on previous costs for similar accidents. Thus, when an accident does occur, an estimate of the final cost can be readily computed using this formula. This estimate can be based on the average costs in the state where the company operates. In approximating actual costs, it is possible to use average costs per medical-only cases, costs for the time away from the job, or daily hospital charges.

An Estimated Cost System

Under an estimated cost system, as soon as an accident occurs, an estimate of the final cost is charged back to the line, which alleviates some of the difficulties in utilizing the dollar, listed above. Because of this advantage, some companies have begun to use this approach. The following original bulletin from the Corporate Safety Department of the St. Regis Paper Company explains such a system.[1]

St. Regis Loss Prevention Bulletin No. 3

SUBJECT: INSTRUCTIONS FOR USING THE INJURY COST SYSTEM

The need in accident prevention to obtain a closer relationship between the actual cost of injuries and the statistical measurement used for evaluation and control has been increasing as compensation costs have climbed. Although St. Regis has long counted *all* disabling injuries, this method of giving each lost-time injury the same weighing regardless of severity and cost has resulted in an incomparable accident frequency rate that does little more than reflect unlike numbers.

Throughout industry, insurance carriers, and associations, a means has been sought to eliminate this shortcoming in accident statistics, with various methods and plans being developed and put to use.

The major weaknesses of the presently used methods are that they cannot be translated into related cost and that the degree of severity or the cost of a fatality or permanent total disability depends on factors that are beyond the control of management. The basis for measuring injuries with an injury cost system is the average national cost as compiled from various statistical sources for each injury and based on the severity of that injury at the time of the first medical evaluation. This information has been used to develop a rate which will reflect both the number of injuries and their cost.

I. PURPOSE

A. An accurate knowledge of facts is essential to an objective evaluation of safety activities.

[1]St. Regis Paper Company, "Instructions for Using the Injury Cost System," New York, 1966.

B. This system has been designed to gather selected information by location to measure performance and progress against prior reporting periods.

II. PROCEDURE

A. The data required for this report should be on a monthly basis.

B. The worker-hours summary accumulates on a monthly and year-to-date basis total worker-hours. It should be noted that this includes *all* hours actually worked.

C. Summary of injuries: The injury cost rate method of reporting and compiling injury information presents an equitable basis for assignment of accountability. It separates the minor injuries or disabilities of short duration from the more serious injuries and weighs each by average cost on a national basis. Both occurrence and severity are reflected in one injury rate. The figures used in this system were developed by using an average cost, based on national surveys for each classification of injury as follows:

1. Medical only	20.00
2. Temporary total	400.00
3. Permanent partial	1600.00
4. Death	16000.00
5. Permanent total	32000.00

By dividing each of the above costs by a common denominator of 400, an average cost factor results, as follows:

1. Medical only	0.05
2. Temporary total	1.00
3. Permanent partial	4.00
4. Death	40.00
5. Permanent total	80.00

This average cost factor gives a weight of 1.00 to a temporary total injury, which is the classification closest to a lost-time injury, and also makes the rate formula more easily handled. At any point, these figures can be converted to average cost by multiplying by 400.

Definition of Terms Used in the Injury Cost System

MEDICAL ONLY is any injury that requires medical treatment other than in-plant first aid and where the injury does not result in temporary total disability.

TEMPORARY TOTAL is any injury in which workmen's compensation weekly benefits are payable but which does not result in permanent impairment. The injured employee must have lost time in excess of your state's waiting period.

WAITING PERIOD is the number of days required by the state in which you operate as a waiting period before an injured employee qualifies for workmen's compensation weekly benefits.

PERMANENT PARTIAL DISABILITY is any injury other than death or permanent total which results in a loss of use of any member or part of a member of the body.

Permanent partial disability also includes permanent impairment of the body or part thereof on either a scheduled or percentage basis.

PERMANENT TOTAL DISABILITY is any injury other than death which permanently and totally prevents an employee from pursuing any gainful occupation.

DEATH is any fatality resulting from a work injury, regardless of the time intervening between the injury and death.

AVERAGE COST FACTOR is the average cost for each of the preceding injuries divided by 400. This is done to give the temporary total injuries a weight of 1 and to facilitate computation of rates.

ADJUSTED INJURY COST FACTOR is the number that results from the average cost factor multiplied by the number of injuries in each type.

TOTAL INJURY COST FACTOR is the total of each of the adjusted injury cost factors. This total multiplied by 400 will result in the total average cost. It is used in the same way that lost-time injuries are used to compute a rate per million worker-hours.

INJURY COST RATE is modified cost per million worker-hours worked. This rate carries the modification of 400 by which the average cost for each injury is divided. The actual average cost for each 1 million hours worked can be computed by multiplying this rate by 400.

Instructions for Completing and Computing Injury Cost Rates

STEP 1: LISTING INJURIES. List all injuries for the month [see Figure 4-2] and for the total year to date under each category: medical only, temporary total, permanent partial, death, and permanent total.

STEP 2: COMPUTE INJURY COST FACTORS. Multiply the number of injuries to date by the average cost factor for each category to get the adjusted injury cost factors. Add for the total injury cost factor.

STEP 3: COMPUTING THE INJURY COST RATE. To compute the injury cost rate, multiply the adjusted injury cost factor by 1,000,000 and then divide by the worker-hours for the year to date. The resulting rate will be average cost of injuries for each million worker-hours of exposure adjusted by 400.

STEP 4: COMPLETING THE REPORT. The section at the extreme left of the summary of injuries is designed to give a ready summary of injury information and to provide ease in compiling data for review. It should be completed by listing all disabling injuries by their codes and the days lost for each injury. The section at the bottom of the report is designed to give a ready summary of accident rate information and to provide ease in compiling data for review. It should be completed by transferring applicable information from the worker-hours, injury, and rate sections of the report. Monthly and year-to-date disabling injury totals should include all temporary totals, permanent partials, and deaths.

The completed report is a tool usable to each location to determine one phase of the cost of doing business. The injury cost rate will serve to emphasize our desire to provide a safe place to work as well as to require safe work behavior.

ST REGIS

MONTHLY SUMMARY OF
PERSONNEL, MAN HOURS AND INJURIES

DISTRIBUTION:
MGR. OF SAFETY (N.Y.O.)
RESIDENT MGR.
REGIONAL IND. REL. MGR.

TOTAL MAN HOURS WORKED (ITEMS MARKED "A")	114,096	(2) 1,138,461

NO. FIRST AID THIS MONTH 16

SUMMARY OF INJURIES

LIST ALL DISABLING INJURIES - CURRENT MONTH		NUMBER THIS MONTH	INJURY TYPE AND AVERAGE COST	TOTAL YEAR TO DATE	AVERAGE COST FACTOR	ADJUSTED INJURY COST FACTOR
ACCIDENT CODE INFORMATION	DAYS LOST	5	MEDICAL ONLY - DOCTOR TREATMENT INVOLVING NO COMPENSABLE DISABILITY	26	X .05 =	1.30
3-55-1301	11	1	TEMPORARY TOTAL - INVOLVING COMPENSATION FOR TEMPORARY DISABILITY	3	X 1.00 =	3.00
			PERMANENT PARTIAL - INVOLVING COMPENSATION FOR PERMANENT DISABILITY	2	X 4.00 =	8.00
			DEATH		X 40.00 =	
			PERMANENT TOTAL		X 80.00 =	
		1.25	TOTAL INJURY COST FACTOR			(1) 12.30

ADJUSTED INJURY COST FACTOR X 1,000,000 / MAN HOURS WORKED = INJURY COST RATE

(1) 12.30 X 1,000,000 / (2) 1,138,461 = 10.80

LOCATION Chicago Bag FISCAL MONTH OF September 19

MAN HOURS CURRENT MONTH	INJURIES THIS MONTH: MEDICAL ONLY	INJURIES THIS MONTH: DISABLING	INJURY COST RATE THIS MONTH	MAN HOURS YEAR TO DATE	INJURIES YEAR TO DATE: MEDICAL ONLY	INJURIES YEAR TO DATE: DISABLING	INJURY COST RATE YEAR TO DATE	ACTUAL AVERAGE COST PER MILLION MAN HOURS
114,096	5	1	10.95	1,138,461	26	5	10.80	$4,320

FIGURE 4-2 Injury cost report, St. Regis Paper Company

The current total injury cost on an average basis reflects the direct cost of our injuries, using a national average for each injury classification. To compute your plant's current injury cost, multiply the number of medical cases by $20, the temporary total injuries by $400, and each other injury class by its cost factor; then add the sum of each classification.

The figures in this system were accurate only for the date of the bulletin (the early 1970s). They must be updated annually to be meaningful. The costing approach is included here to explain the concept, and the reader should use only updated costing figures when utilizing the concept.

ACCOUNTABILITY FOR ACTIVITIES

Figure 4-3 lists some of the items against which management might measure the line organization to determine what they are *doing* to prevent accidents from occurring. This measurement is *accountability for activities.* It is perhaps more important than the measurement of results because it measures line competence in controlling losses *before* accidents happen.

Management can measure line supervisors to see whether they are utilizing such techniques of accident control as toolbox meetings, job hazard analyses, inspections, accident investigations, incident reports, safety committees, and safety meetings. Management may require line supervisors to submit activity reports. All these activities are known to be effective in safety. When management measures them, it is setting up a system of accountability for activities.

Measurement Tools

Also listed in Figure 4-3 are techniques which management might use in measuring line safety activities. Many new statistical controls are now being used—patterned, for the most part, on the quality control systems used in industry so successfully for many years. Safe-T-Scores and statistical controls tell us when the record is significantly different, that is, when things are beginning to go sour.

Critical incident techniques tell us where accidents can be expected to occur in the future—not where they have happened in the past.

Safety Sampling

Safety sampling is a very useful technique in industrial safety. It has the potential of telling management which line supervisors are doing their job in safety and which are not. Safety sampling is a method of systematically observing workers in order to determine what unsafe acts they are performing and how often. The results of these observations are then used to measure the effectiveness of line safety activities.

In utilizing safety sampling, first a code is prepared. The most common unsafe acts are listed on a form, and each is given a code number. Next, an observer takes a sample by walking rapidly through the operation and observing each employee quickly. An immediate decision is made concerning whether

ACCOUNTABILITY FOR ACTIVITIES

Management measures what supervisor is doing

(1) Safety meetings held
(2) Tool box meetings
(3) Activity reports on safety
(4) Inspection results
(5) Accident investigation
(6) Incident reports
(7) Job hazard analyses

Management measuring tools

(1) Safety sampling
(2) Statistical controls
(3) Critical incident techniques
(4) Safe-T-Scores

FIGURE 4-3

each employee is working safely. If it is noted that an employee is working safely, a check is made on a theater counter, indicating one safe observation. If an employee is seen to be working unsafely, the observer records this as one unsafe act and notes the code number of the act.

The third step is to validate the sample statistically to determine whether there are enough observations to constitute a representative sample. The fourth and final step is to prepare a report for management. This shows each supervisor's rating, expressed as a percentage of safe to unsafe acts. This can be compared with the supervisor's past record and with the records of other departments. From this, management can judge line performance and take whatever action is necessary. Safety sampling is measuring supervisory performance. It is fixing accountability on the basis of what is happening today, rather than waiting for accidents to occur. Safety sampling is mentioned here as one example of fixing accountability on the line. (Further details are discussed in Chapter 6.) It is an excellent indicator of supervisory performance and also an excellent motivator of supervisory personnel.

SCRAPE

SCRAPE is a systematic method of measuring the accident prevention effort. As we have seen, most companies are set up to measure accountability through analysis of results. Monthly accident reports at most plants suggest that the supervisor should be judged by the number and cost of accidents that occur in his or her department. We should judge line supervisors by what they *do* to control losses. SCRAPE is one way of doing this. It is a simple method—as simple as deciding what we want supervisors to do and then measuring to see that they do it.

The SCRAPE rate indicates the amount of work done by a supervisor and by the company to prevent accidents in a given period. Its purpose is to provide a tool for management which shows—before an accident happens—whether positive means are being used regularly to control losses.

The first step in SCRAPE is to determine specifically what we wish the line manager to do in safety. Normally this falls into the categories of (1) making physical inspections of the department, (2) training or coaching people, (3) investigating accidents, (4) attending meetings with the boss, (5) establishing safety contacts with employees, and (6) orienting new people.

With SCRAPE, management selects which of these things it wants supervisors to do and then determines their relative importance by assigning values to each. Let us suppose that management believes that the six items above are the things it wants supervisors to do; that items 1 and 2, inspections and training, are the most immediately important, followed by accident investigations and individual employee contacts; and that attending meetings and orienting new people are relatively less important at this time. Management might then assign these values:

Item	*Points*
Departmental inspections	25
Training or coaching (e.g., five-minute safety talks)	25
Accident investigations	20
Individual contacts	20
Meetings	5
Orientation	5
Total	100

Depending on management's desires, the point values can be increased or decreased for each item. They should, however, total 100.

Every week all supervisors fill out a small form (see Figure 4-4), indicating their activities for the week. Management, on the basis of these forms, spot-checks in all six areas, notes the quality of the work done, and rates the accident prevention effort by assigning points between 0 and the maximum.

For example, in department A the supervisor conducts an inspection and makes six corrections. The safety director later inspects and finds good physical conditions. Supervisor A rates the maximum of 25 points.

In department B the supervisor states that an inspection was made but that there were no corrections. The plantwide inspection, however, indicates that much improvement is needed. Supervisor B might get only 5 points; the inspection was poor.

In department C there were five accidents. Only one individual lost time, and supervisor C turned in only one investigation. Supervisor C thus gets only 5 points for the effort in this area.

In department D there are 43 employees, but only three were individually contacted during the week. This might also be worth only 5 points.

Department ____________ Week of ____________ Points

(1) Inspection made on ____________ # corrections ______ ______
(2) 5-minute safety talk on ____________ # present ______ ______
(3) #accidents ____________ # investigated ______

Corrections ____________
____________ ______

(4) Individual contacts:
Names ____________
____
____
____________ ______

(5) Management meeting attended on ____________ ______
(6) New men (names): Oriented on (dates):
____________ ____________
____________ ____________
____________ ____________ ______

FIGURE 4-4 SCRAPE activity report form

Management should decide on relative values by setting maximum points and should also set the ground rules concerning how maximum points can be obtained. Each week a report is issued (see Figures 4-5 and 4-6).

SCRAPE can provide management with information on the company's accident prevention performance. It measures safety activity—not a lack of safety. It measures before the accident happens—not after. Most important, it makes management define what it wants from supervisors in safety and then measures to see that it is achieving what it wants.

ACCOUNTABILITY SYSTEMS USED IN CONSTRUCTION

Too many textbooks and articles on safety discuss systems that are usable in plants but have no real application in the construction industry. The idea of fixing accountability for safety to line management certainly applies in construction. Some good progress has been made in this area in the construction industry.

Management principles apply to contractors just as they do to plants. The principle of accountability holds. It is perhaps even more important in construction because of the physical separation of management and job supervision.

Below we discuss two methods of holding supervisors accountable that have been used in the construction industry. The two systems were devised by Walt Willard, formerly with Industrial Indemnity Company of San Francisco.

An Adjusted Severity Rating Plan

This plan consists in recording a combination of the frequency and severity rates of all job accidents that result in injuries serious enough to require outside medical treatment. The results which represent each job superintendent's monthly record are obtained by the following equation:

Week of							
	Activity						
Department	Inspect (25)	5-min talks (25)	Acc inv (20)	Ind cont (20)	Meet atten (5)	Orient (5)	Total rate (100)
Average							

FIGURE 4-5 SCRAPE weekly report

Week of ______							
	Activity						
Department	Inspect (25)	5-min talks (25)	Acc inv (20)	Ind cont (20)	Meet atten (5)	Orient (5)	Total rate (100)
A	25	15	20	15	5	5	85
B	5	10	20	5	5	5	50
C	25	10	5	5	5	5	55
D	15	25	20	20	-	5	85
E	10	5	-	-	5	-	20
F	20	20	15	5	-	-	60
Average	17	14	13	8	3	3	58

FIGURE 4-6 SCRAPE weekly report

$$\frac{\text{Time lost by all job injuries} \times 1{,}000}{\text{Worker-hours of exposure}}$$

Time lost by all injuries represents four hours for each minor injury and a minimum of 40 hours for a serious or lost-time injury or fatality. A minor injury is interpreted as one requiring outside medical treatment but not so serious as to prevent the employee from returning to work either the same day or the next regular working day. A lost-time job injury is interpreted as one that prevents the injured employee from returning to work; an eight-hour penalty charge for each day, up to a maximum of 40 hours for five days, is assessed. The superintendent with the lowest-numbered record would have the best record.

SUPERINTENDENT A = average 20 employees
2 minor injuries requiring outside medical attention = 8 hours lost
20 employees = 3,200 hours exposure per month
Time lost × 1,000 divided by worker-hours = result

8 × 1,000 = 8,000

8,000 ÷ 3,200 = 2.5 adjusted result

SUPERINTENDENT B = average 10 employees
2 minor injuries requiring outside medical attention = 8 hours lost
10 employees = 1,600 hours exposure per month

8 × 1,000 = 8,000

8,000 ÷ 1,600 = 5.0 adjusted result

SUPERINTENDENT C = average 50 employees
1 major injury = 40 hours lost
1 minor injury requiring outside medical attention = 4 hours lost

TOTAL 44 hours lost

$44 \times 1{,}000 = 44{,}000$

50 employees = 8,000 hours exposure per month

$44{,}000 \div 8{,}000 = 5.5$ adjusted result

Superintendent A has the best record with an adjusted rate of 2.5.

The Jobsite Survey by Management

At least once each month management makes a jobsite survey covering the 11 factors specified below. A rating is made of each factor, and a composite rate is computed. This gives a "reading" on the safety activities of each job supervisor. This evaluation covers 11 areas, as follows:

1. Safety meetings. Superintendents conduct short safety meetings on the job at least once every 10 working days. Written verification of these meetings is entered on the daily job log.
2. Job inspection. The job superintendent makes at least one safety tour of the job each day. Deficiencies are noted and corrected.
3. New-employee orientation. Every new employee is given a copy of "Job Safety Rules" and is directed to read the posted "Safe Practice and Procedure Code." The new worker is observed for a reasonable period to make certain that he or she is working in a safe manner.
4. Adjusted severity rates. (See the discussion above.) Each job is compared individually with the company average. Superintendents compute the rates and enter them on the job log monthly.
5. Public liability. Every reasonable precaution is taken to protect the public from harm on or around the projects. Procedures must be established that will provide protection 24 hours a day, every day, during the contract period.
6. Fire protection. The phone number of the local fire department is posted. Sufficient fire extinguishers are readily available. The jobsite is checked for potential fire hazards each day before the job is secured.
7. Housekeeping. Good housekeeping and fire protection go together. This is particularly true of high-rise structures. Cleanup is scheduled as an integral part of each operation. Subcontractors are advised that they are expected to maintain good housekeeping practices.
8. Hard hats. The state construction safety orders require that all workers who are subjected to the hazard of falling or flying material wear hard hats.

JOB SITE SURVEY

Job No. ______________ Superintendent ______________ Date ________

	Maximum points allowable per month	Points
(1) Safety meetings held and recorded every 100 working days? No - 0 Yes - 5 Subs attend. -15		
(2) Job inspection made Every day - 10 Once a week - 5		
(3) New employee orientation? No - 0 Yes - 5		
(4) Adjusted severity rating Under company coverage - 10 Over - 0		
(5) Public liability exposures 0 through 5		
(6) Fire prevention and control 0 through 5		
(7) Housekeeping Poor - 0 Fair - 10 Excellent - 15		
(8) Hard Hats Deduct two points for each employee not wearing a hard hat at the time the survey was made		
(9) Floor, roof, and wall openings guarded? Yes - 10 Deduct 5 points for each violation		
(10) Ladders, stairways, scaffolds in good order? Yes - 10 Deduct 5 points for each violation		
(11) Subcontractor safety orders violated? No - 5 Deduct 1 point for each violation		
Total		

FIGURE 4-7 Jobsite survey form

9. Floor, roof, and wall openings. These requirements are covered by state safety orders. Subcontractors are expected to fill them in as soon as possible.

10. Ladders, stairways, and scaffolding. Ladders and scaffolds are covered by the state safety orders. Stairways and landings must be free of debris, ladders must be properly placed, and the exit areas at the top and bottom must be clear. All scaffolding over 7½ feet high should have the required back rails.

11. Subcontractors must comply with all applicable state construction safety orders.

The evaluation of these 11 factors is made on the form shown in Figure 4-7. Every month each job superintendent is evaluated and measured against the company average.

REFERENCES

Cook, Ken: "Safety Sampling," Ken Cook Lectron Company, Milwaukee, Wis., 1963.

Martin, J. A: "Large Plant Safety Program Management," *Journal of the ASSE,* May 1963.

Pollina, Vincent: "Safety Sampling," *Journal of the ASSE,* August 1962.

Tarrants, William: "Applying Measurement Concepts to the Appraisal of Safety Performance," *Journal of the ASSE,* May 1965.

PART THREE

MEASURING SAFETY PERFORMANCE

In Chapter 4 we discussed the concept of accountability and defined it as "the fact of active measurement by management to ensure compliance with its will." This part is about that "active measurement," and we shall look at measurement in some detail. In Chapter 5 we shall see what makes a good measure and what makes a poor measure, and in doing this we shall develop some criteria for measurement. Obviously, these criteria would be different at different levels of the organization. What properly measures the performance of a first-line supervisor might not measure a company's performance adequately. We shall develop criteria for use at different levels of the organization: at the first-line supervisory level, at the upper and middle managerial levels, and at the corporate level. Then we shall attempt to develop measures for use at the national level.

In Chapters 6 to 9 measures will be discussed that are in use in industry, and we shall see how they meet the criteria established. Finally, Chapter 10 examines, as a transition to Part 4, the relationship between the measuring sticks used and the motivation produced in the people being measured. I believe that the terms "measurement" and "motivation" become almost synonymous at many levels of management.

5

CRITERIA FOR SAFETY MEASUREMENT

In Chapter 4 the concept of measurement was introduced as part of the discussion of accountability. Perhaps we should not think of the two as being separate since in practice the terms are almost synonymous. To hold someone accountable, we must know whether he or she is performing well; we must measure that person's performance. Without measurement, accountability becomes an empty and meaningless concept.

Thus, starting with what we propose as the single most important factor in getting good line safety performance—accountability—we find that we are really talking about ways to measure the line manager better. And measurement has been our downfall in safety for years—at least at the corporate and national levels. Measurement has been discussed over and over again at these levels, and we have yet to come up with a meaningful measure of safety performance. Perhaps our inability to create these needed measures is one reason for our lack of good safety performance. We have not, however, devoted as much time to good measures of lower management's safety performance, and this might also account for our inability to get good safety performance from line organizations.

Chapter 10 discusses the notion that, for the line manager, "to measure is to motivate." Although this statement might have sounded a little ridiculous 20 years ago, I believe that it expresses a profound truth, at least in terms of the safety performance of line organizations. Managers react to the measures used by the boss; they perceive a task to be important only when the boss thinks it is worth measuring.

THE SAFETY PROBLEM

Having perceived the importance of measurement in obtaining good safety performance, we then hit our biggest snag: What shall we measure? Should we measure our failures as demonstrated by accidents that have occurred in the past? If this is in fact a good measure, as has historically been believed (for that is what we usually measure), then what level of failure should be measured? We can measure the level of failure we call "fatalities." Fatalities are used to measure our national highway traffic safety endeavors. Is the measure of fatalities, then, a "good" measure? Obviously, we cannot answer this question until we examine the size of the unit being measured. Fatalities could be a good measure if we are assessing the national traffic safety picture, but it would be a little ridiculous in the case of a supervisor of 10 factory workers. Such a supervisor might well do absolutely nothing to promote safety and still never experience a fatality in his or her department. Obviously, measuring fatalities would make little sense in this case.

Unfortunately, our traditional frequency rate is not much better than fatalities in the example above when we use it as a measure of supervisory performance in safety. It measures a level of failure somewhat less than a fatality (an injury serious enough to result in a specified amount of time lost from work), but the fact remains that a supervisor of 10 workers can do absolutely nothing for a year and attain a zero frequency rate with only a small bit of luck. By rewarding such a supervisor, we are actually reinforcing nonperformance in safety. He or she learns that it is not necessary to do anything in order to get a reward. While this may sound a little ridiculous, it accurately describes what is going on in many safety programs today.

If fatalities (or frequency rates) are a poor measure of supervisory performance, what is a good measure? Or, more important, what is wrong with fatalities as a measure? Perhaps measuring our failures is not the best approach to use in judging safety performance. After all, this is not the way we measure people in other aspects of their jobs. We do not, for instance, measure line managers by the number of parts the people in their departments failed to make yesterday. And we do not measure the worth of salespeople by the number of sales they did not make. Rather, in cases like these we decide what performances we want, and then we measure to see whether we are getting them.

What would be a good measure of supervisory safety performance? More important, what set of criteria can we develop for measuring supervisory safety performance? Or the safety performance of the corporation? Or our national traffic safety performance? Or anything else related to safety?

Even a brief look at the problem of measurement shows us that we need different measures for different levels in an organization, for different functions, and perhaps even for different managers. What is a good measure for one supervisor of 10 people may not be a good measure for another, much less for a plant superintendent or the general manager of seven plants and 10,000 people. What might be a good measure for the supervisor of a foundry cleaning room may be inappropriate for use in judging the effectiveness of OSHA.

The purpose of this chapter is merely to suggest some criteria for different levels and for different functions. The following four chapters suggest measures for meeting these criteria.

CRITERIA

The development of criteria for good safety measures is certainly not an easy task. We have been grappling with it for years, and the most noted theorists and scholars in safety have been writing on the problem since the late 1950s. From their writings (notably those of Dr. Tarrants, Dr. Rockwell, and Dr. Grimaldi), general criteria for safety measures can be distilled.

We need more than a list of general criteria, however. For instance, such a list usually includes an item called "statistical reliability," which has to do with whether a measure tends to fluctuate wildly when there has not been much change in the system being measured. The criterion of statistical reliability would obviously be useless for measuring a first-line supervisor's safety performance. A supervisor who does nothing for safety (zero performance level) could have a perfect record or a miserable record. At the supervisory level, statistical reliability is next to impossible to achieve with any measure—the data base is simply too small.

Other examples could be given, but perhaps none are necessary. Different criteria, and therefore different measures, seem to be needed for different organizational levels and perhaps for different functions.

The following sections describe criteria for different levels and different functions. These were developed by relating what has been written in the field since the 1950s to different organizational levels. Much of the thinking in these sections has come from discussions held with, and papers written by, graduate students in safety management at the University of Arizona. You, the reader, may agree or disagree with some of the items included in each list of criteria. The important thing is that you do in fact decide on some criteria of your own that you believe in, for without criteria we cannot do a good job of developing better measures.

Safety-Measurement Criteria for Employees

Few measures of safety performance are used at the lowest organizational level, that of the employee. Those which are used are essentially individual assessments of performance. They seem to be supervisory techniques, rather than measures. Nonetheless, we can consider a few possible criteria for a measure of employee safety performance:

1. It should be so constructed that it can be used to affect the employee's rewards (appraisal, promotions, bonuses, etc.).
2. It should be constructed in such a way that it recognizes or can be used to recognize safe performance (rather than unsafe performance).
3. If possible, it should be self-monitoring. (See Chapter 13 for a discussion of this aspect of measures as it relates to employee motivation.)

4. It should be motivating to the employee.

This list is shorter than those suggested for other levels, but it should give the reader an idea of what safety-measurement criteria are.

Safety-Measurement Criteria for First-Line Supervisors

Measurement is more crucial at this level, and the measure (which is the motivator here) must do many more things than at the employee level. The following are a few criteria:

1. It should be flexible to encompass individual managerial styles and different strategies that supervisors use to get things done.
2. It should give swift and constant feedback.
3. It should be able to be used to judge promotability.
4. It should get the supervisor's attention.
5. It should measure the presence of safety activity, not only its absence (as indicated by accidents).
6. It should be sensitive enough to indicate when effort has slowed.
7. It should be able to provide an alert, showing that something is wrong.
8. It should be understandable to those at both the supervisory and upper-management levels.
9. It should be able to be used to provide recognition of supervisors' efforts.
10. It should allow for creativeness.
11. It should be valid; that is, it should measure what it was intended to measure. If you want performances from supervisors such as accident investigations, inspections, and training, etc., the measure should tell whether you are getting these performances. It should not measure only failures (accidents) as an indication of whether you are getting the desired performances.
12. It should be mainly performance-oriented.
13. Insofar as possible, it should be self-monitoring.
14. It should be meaningful.

Obviously, no one measure will meet all the above criteria. Thus there will have to be some trade-offs, and it will be necessary to devise and use measures that meet as many of the criteria as possible. In the next chapter we shall examine some traditional and nontraditional measures to see how well they meet these criteria.

Safety-Measurement Criteria for Middle and Upper Management

Many of the same criteria apply at the upper levels of management as at the lower levels, although some new ones become important. A measure of performance at the upper- and middle-management level, should meet the following criteria:

1. It should be flexible enough to allow for individual managerial styles and strategies.
2. It should be capable of giving swift and constant feedback.
3. It should be able to be used in judging promotability.
4. It should get the attention of those in middle and upper management.
5. It should measure the presence, as well as the absence, of safety.
6. It should be sensitive to change (ablc to alert management to new situations and problems).
7. It should be understandable to both the middle manager and those in top management.
8. It should be built in such a way that it can offer recognition for performance.
9. It should allow for creativeness.
10. It should be valid.
11. It should be both performance- and results-oriented.
12. It should be meaningful.

In Chapter 7 we shall examine current and proposed measurements and see how well they meet these criteria.

Safety-Measurement Criteria for Corporate Management

The measure used at this level indicates how well the company is doing. It is used only internally, within the organization. It can indicate the progress of the entire organization and could be construed as constituting a judgment of the president and controlling officers. Such a measure is not intended for use in comparing the performance of one corporation with that of another (different criteria will be suggested for this type of measure). The following are the most important criteria for a corporate measure of safety performance for internal uses only:

1. It should be valid; that is, it should measure what it was designed to measure.
2. It should be statistically reliable; it should not fluctuate without reason.
3. It should be objective.
4. It should be meaningful to management.
5. It should be quantifiable.
6. It should be stable.
7. It should be sensitive to changes and problems.
8. It should ensure input integrity.
9. It should be primarily results-oriented.
10. It should be able to be computerized.
11. It should point up weaknesses in the system and thus make it possible to take preventive action.

Since we are dealing here with larger numbers of people, we can demand certain things of this measure that we could not expect of others: statistical reliability, quantifiability, input integrity, and the ability to be computerized, for example. These are ideal criteria for a measure, and at this level of the organization we can ask for them. In Chapter 8 we shall look at measures in the light of these criteria.

Safety-Measurement Criteria at the National Level

Here we are discussing measures of corporate safety performance that can be used to compare one company's progress with that of another and with the national average. Such a measure should meet these criteria:

1. It should be valid.
2. It should be statistically reliable.
3. It should be objective.
4. It should be meaningful to anyone involved in its use.
5. It should be quantifiable.

6. It should be stable.
7. It should be understandable to the layman.
8. It should be results-oriented only.
9. It should be able to be computerized.

A measure used to judge our collective progress at the national level would have to meet the following criteria:

1. It should be valid.
2. It should be statistically reliable.
3. It should ensure input integrity.
4. It should be totally results-oriented.
5. It should be understandable to all.

Safety-Measurement Criteria at the Staff Level

A measure of staff safety performance should meet the following criteria:

1. It should be valid.
2. It should be understandable to management.

Additional Criteria

The following additional criteria for a safety measure should be added to all the above lists:

1. It should have a good cost-benefit ratio.
2. It should be adminstratively feasible.
3. It should be practical.

The criteria listed and discussed in this chapter reflect the thinking of only a few people and are certainly not purported to be universally accepted. However, we have to start somewhere. Unless we define our criteria, we cannot tell whether the measures we devise are worthwhile.

REFERENCES

Grimaldi, John V.: "The Measurement of Safety Engineering Performance," *The Journal of Safety Research,* September 1970.

Rockwell, Thomas, and **Vivek Bhise:** "Two Approaches to a Non-Accident Measure for Continuous Assessment of Safety Performance," *The Journal of Safety Research,* September 1970.

Tarrants, William: "A Definition of the Safety Measurement Problem," *The Journal of Safety Research,* September 1970.

6

MEASURING THE LINE MANAGER'S PERFORMANCE

The principle of accountability cannot be separated from the techniques of measurement. Measurements are made for the purpose of fixing accountability—accountability without measurement is meaningless.

In this chapter we discuss measurement techniques that fix accountability to line management. Several of the more common tools of safety measurement will be discussed: inspection, accident investigation, record keeping, and statistical control techniques. These tools will be examined as they relate to the principle of accountability—not just as tools in themselves. They will be surveyed from some new angles—rather than from the same perspective in which they are usually seen. In the past, safety professionals have tended to use inspection for the purpose of seeking out hazards. They have used accident investigation for the purpose of identifying an unsafe act or an unsafe condition, and they have used record keeping only to compute frequency and severity rates.

Inspections have been used to spot conditions but seldom to spot acts. Investigations have been used to unearth symptoms rather than causes. Records have been used to tabulate accident types, accident agencies, injury types, etc., and not accident causes. This chapter discusses how we might better use these tools for what should be their primary purposes: measuring line safety performance, spotting acts as well as conditions, and unearthing causes as well as symptoms. We shall look at the first-line measurement problem in total and discuss other means of measuring performance at this level.

INSPECTION

Inspection was and is one of the primary tools of the safety specialist. Before 1931 it was virtually the only tool, and from 1931 to 1945 it was still the one most used. Until 1960 it was the primary tool of many outside service agencies, and even today it remains the primary (and sometimes the only) tool of some safety professionals.

Today, however, I believe that one key question should be asked by every safety specialist engaged in inspection: "Why am I inspecting?" The answers to that question dictate how, when, and where to inspect. For instance, if we are inspecting in order to unearth physical hazards only, we will look only at things. If, however, we are inspecting in order to pinpoint both physical hazards and unsafe acts, we will also look at people. Unfortunately, most inspections today are of the former kind, rather than the latter.

If our primary intent is to detect hazards that we have not seen before, we inspect differently from the way we do if our primary interest is in checking on the inspections the department supervisor has made. If our intent is to detect hazards only, we can immediately have them corrected by going directly to the maintenance department and reporting any deficiencies. If our intent is to audit the supervisor's inspection, we will use what we find to instruct and coach the supervisor so that future inspections will be improved.

Many articles have been written on safety inspections, and many have asked the question: Why inspect? Some typical answers have been:

1. To check the results against the plan
2. To reawaken interest in safety
3. To reevaluate safety standards
4. To teach safety by example
5. To display the supervisor's sincerity about safety
6. To detect and reactivate unfinished business
7. To collect data for meetings
8. To note and act upon unsafe behavior trends
9. To reach firsthand agreement with the responsible parties
10. To improve safety standards
11. To check new facilities
12. To solicit the supervisor's help
13. To spot unsafe conditions

The single most important reason for making inspections is seldom mentioned. It is:

14. To measure the supervisor's performance in safety

Perhaps if the line manager felt that this was the primary purpose of management's inspection, he or she might do a better job of making sure that nothing amiss could be found in the department.

If inspection is used as a measurement tool of accountability, the line manager will start inspecting more often to ensure that conditions remain safe and that fewer unsafe acts are being performed. The supervisor will not wait until the safety specialist comes around to do the inspection job.

Who Is Responsible for Inspection?

It is generally agreed that the responsibility for conditions and for people is the line supervisor's. Thus responsibility for the primary safety inspection must also be assigned to the supervisor. By "primary safety inspection" we mean the inspection intended to locate hazards. Any inspections performed by staff specialists then should be only for the purpose of auditing the supervisor's effectiveness. Hence the results of our inspection become a direct measurement of safety performance—effectiveness.

Symptoms of Unsafe Conditions

"An unsafe act, an unsafe condition, and an accident are all symptoms of something wrong in the management system." This principle, expressed in Chapter 2, ought to be constantly in our minds as we inspect. We ought to look behind the acts we see when inspecting, and behind the conditions we find, and ask: Why are these here? The answer to this question may lead us back to the department supervisor—it may even lead us to some other system weakness within the company—but the question should be explored fully and answered.

For instance, if an unsafe ladder is discovered, the inspector should immediately ask such questions as: Why is this ladder here? Why was it not uncovered by our ladder inspection procedure? Why did the line supervisor allow it to remain here? Answers to questions such as these begin to get at the true causes of accidents.

Recommendations and Reports

Instead of submitting lists of recommendations in their reports, which is still typically done today (see Figure 6-1), safety professionals might submit reports showing suggested changes in management procedures (Figure 6-2). The report shown in Figure 6-1 is all too typical of those submitted today. There is no search for the causes of the unsafe conditions; the inspection is designed only to remove symptoms. The report in Figure 6-2 represents a better approach to the safety specialist's inspection. Another approach might be to have the report show only the results of the specialist's discussion or the results of the coaching session with the supervisor involved.

The Primary Inspection Checklist

The line supervisor is often provided with a checklist to use for the primary inspection, such as that shown in Figure 6-3. This approach does not encourage

INSPECTION REPORT

By John Jones, S.D. Date 1/16 D. C. Anderson

General conditions: J. C. Hansen

Housekeeping: Fair, some places need attention

Equipment: Generally good

Hand tools: Fair, some need repair

Lighting: Good

Ventilation: Good except in building 4

Floors: Good

Guards: See below

Recommendations

Building 4, third floor - Change drill placement to avoid crowding.
Guard power press No.413.

Building 5, first floor - Change control switch on milling machine.
Adjust tool nuts on all grinders so rest is at 1/8 inch from the wheel.
Advise against hanging goggles on the machines for general use.

FIGURE 6-1 Typical inspection report

an effort to trace back the symptoms to their true causes. This type of checklist should be reworded into a form (see Figure 6-4) that requires determination of some of the causes of the symptoms that have been unearthed.

RECORDS

Accident records have always been a key tool of the safety professional, and they are still important today. Here again, however, a key question should be asked: "Why are we keeping records?" The answer to this question dictates the kind of records that should be kept. Much has been written over the years on accident record keeping, but, again, the reasons for keeping the records are seldom identified.

Our disscussion of records in this chapter is based on the premise that the primary answer to the above question is: "We keep records in order to measure supervisors." Two different categories of accident records are generally kept in industry: (1) accident investigation records and (2) injury records. (See

"Accountability for Results," in Chapter 4, and "Failure Measures," later in this chapter.)

Accident Investigation Records

The primary accident investigation function has always been the supervisor's. Usually management provides the supervisor with a simple form on which to record the results of the investigation (see Figure 6-5).

Here again we ask the supervisor to investigate thoroughly and determine only one "cause" for the accident. This violates our multiple-causation principle. Asking the supervisor to identify only one act or condition violates the principle that the act and the condition are merely *symptoms* of the accident, not the cause.

The tools that we give supervisors ought to lead them to determine some of the many underlying causes. It is proper for line supervisors to investigate. But we should allow them to determine what really happened—and not tell them to

DEPARTMENT AUDIT

Department 43 Supervisor Bill Persons
Audit by John Jones, S.D. Date 1/16 cc: D. C. Anderson
J. C. Hansen

Appraisal of supervisor's inspection performance

Bill does an adequate job of inspecting the department, but doesn't seem to get to it often enough.

Symptoms noted

Conditions	Discussed with supervisor?	Cause found?	Disposition
Missing press guard	Yes	Mainly training	Bill to handle
Defective ladder	Yes	SOP is weak	I will handle
Acts			
Operating without guard	Yes	SOP is weak	Bill will re-write the SOP

Other

Suggestions Bill has agreed to a weekly scheduled inspection. I will follow up in one month.

FIGURE 6-2 Suggested inspection report

SUPERVISOR'S INSPECTION FORM

Name ______________________ Date ______________

Item:	Good	Poor	Disposition
Housekeeping			
Aisles			
Piling			
Floor surfaces			
Tools			
Condition			
Grounding			
Guards			
Personal protection			
Miscellaneous			
Ladders			
Slings			

FIGURE 6-3 Typical supervisor's inspection form

stop thinking about it after identifying one contributory act or one condition. Perhaps the form shown in Figure 6-6 would be helpful.

To tie in the "measurement" or "accountability" idea to accident investigation, it would seem logical that:

1. All accidents, not merely lost-time accidents, should be investigated by the supervisor.

2. Management should receive the investigation form. (It must be transmitted up the line.)

3. At least five possible causes should be identified on each investigation.

4. At least two measures should be taken to prevent a recurrence.

As is true of the inspection tool, the primary investigation must be the responsibility of line management; however, in certain instances it is certainly desirable for staff safety people to make a further investigation to determine

SUPERVISOR'S INSPECTION FORM

Name ______________ Date ______________

Symptom noted Act/Condition/Problem	Causes Why – What's Wrong	Corrections made or suggested By you – By others

FIGURE 6-4 Suggested supervisor's inspection form

SUPERVISOR'S REPORT OF INJURY

Name of injured ______________

Injury date ______________ Time ______ A.M.–P.M.

Did injured return to work ? ______ Time ______ A.M.–P.M.

Witnesses ______________

Nature of injury ______________

Where and how did the accident occur ? ______________

Unsafe act or condition ______________

Measures taken in preventing a similar type of accident ______________

FIGURE 6-5 Typical accident investigation report

SUPERVISOR'S REPORT OF INJURY

Name of injured ____________________

Injury date ____________ Time ________ A.M.–P.M.

Did injured return to work ? ________ Time ________ A.M.–P.M.

Witnesses ____________________

____

Nature of injury ____________________

____

Where and how did the accident occur ? ____________________

____

____

Identify:

Acts and conditions	Possible causes

Measures taken in preventing a similar type of accident
(List on the reverse side)

Supervisor's signature ____________ Department ____________

FIGURE 6-6 Suggested accident investigation report

causes. Usually more detailed digging can be done by the staff when it seems important that underlying causes be identified.

Some organizations routinely specify that the safety department will investigate all serious injuries; others leave the choice to the specialist. Some organizations have set up elaborate plans for detailed analyses of operational errors (and even potential errors).

PERFORMANCE MEASURES

Thus we start our examination of supervisory measurement by looking at the things the supervisor does to get results and by determining whether the supervisor actually does them. We find ourselves thinking in terms of measures like these:

1. In the area of inspections
   - *a.* How many have been made?
   - *b.* How many unsafe conditions were found?
   - *c.* How many corrections of conditions were made?
   - *d.* How many corrections of conditions were suggested?
   - *e.* Were behaviors observed?
   - *f.* How many poor performances were noted?
   - *g.* How many corrections of performance were made?
   - *h.* How many corrections of performance were suggested?
   - *i.* Were system weaknesses found? How many?
   - *j.* How many system changes were made?
   - *k.* How many system changes were suggested?
2. In the area of accident investigations
   - *a.* How many were made (in relation to the number of accidents)?
   - *b.* Were they made on time?
   - *c.* How many root causes were found (in the management system)?
   - *d.* How many corrections were made?
   - *e.* How many corrections were suggested?
   - *f.* What was the quality of the corrections?
3. In the area of training
   - *a.* How many new employees were trained in safety?
   - *b.* How many old employees were trained in safety?
   - *c.* How many five-minute safety talks were given?
   - *d.* How many training sessions were attended?
   - *e.* What scores were made on the tests given during training?
   - *f.* What improvements resulted from the training?
4. In the area of motivation
   - *a.* How many employees were individually contacted?
   - *b.* How many positive or negative "strokes" (reinforcements) were given?

   c. How many posters were used?

   d. What other media were utilized?

5. In other areas

   a. How many JSAs (job safety analyses) were done?

   b. How many safety samples were taken?

   c. How many JSOs (job safety observations) were made?

   d. How many hazard hunts have been submitted?

In short, performance measures are simply counts of the supervisor's performances. They meet our criteria well. For instance:

1. They are flexible and take into account individual supervisory styles. We do not have to use the same measure for all supervisors. We can allow each supervisor to select performance and levels of performance and then measure these performances and levels. Performance measures are excellent for use in MBO (management by objectives) approaches.
2. They give swift feedback, since most require supervisors to report their level of performance to the boss. (They are also often self-monitoring.)
3. They all measure the presence, rather than the absence, of safety.
4. They usually are simple and thus administratively feasible.
5. They meet most of the other criteria for measures of safety discussed in Chapter 5.

RESULTS MEASURES

Results measures can be used either before an accident occurs or afterward. After-the-accident measures might be considered measures of our failures.

Failure Measures

Failure measures tend to be generated from our injury record system. Injury records should be designed so that they measure the line manager. In Chapter 4 we talked about accountability for results. Almost anything we do in this area must stem from our injury record system.

In order that the injury records can be used for measuring the results of the line manager's safety performance, they should be set up so that:

1. They are broken down by supervisor (by department).
2. They give some insight into the nature and causes of accidents.

3. They are expressed eventually in terms of dollars by department (by supervisor).

4. They conform to any legal and insurance requirements.

Beyond these broad outlines, each company can devise any system that seems right for it. The "dollar" criteria is included because of the belief that a dollar measuring stick is much more meaningful to line personnel than any safety specialist's measuring stick (such as our frequency and severity rates).

Many companies today have elaborate and successful record systems that do not measure up to these criteria. Obviously, if they serve their own intended purpose, this is the most important factor.

With failure measures we can count incidents, accidents, or injuries, and we can count these at various levels of severity. For example, we can count only fatalities, only lost-time injuries at the level of seven or more days lost (which is when compensation starts in some states), only those resulting in one or more days lost (as in our traditional severity rate), only those requiring the attention of a doctor, or perhaps only cases requiring first aid. We can even go all the way and start counting all close calls.

What we choose to count will change our results markedly. For lower-level management, it is best to use a measure that gives us lots of numbers. For a supervisor of 10 people, a measure of close calls as well as injuries makes more statistical sense and is more meaningful than one that counts only fatalities or only lost-time injuries.

Before-the-Fact Measures

Before-the-fact measures measure the results of supervisor action before an accident occurs. For instance, we use inspections in this way, as when a periodic inspection is made of a supervisor's work area to measure how well he or she is maintaining physical conditions there. This is a measure of whether things are wrong and, if so, how many things are wrong, We can also measure how well a supervisor gets through to the people in the department by measuring these people's work behavior. Safety sampling, which is discussed later, is used for such measurements.

Some results measures can meet many of our criteria also. For instance:

1. They can be constructed to give swift feedback.

2. They can be used for promotion purposes.

3. They can attract attention.

4. They can measure the presence, not the absence, of safety.

5. They can be sensitive to change.

6. They can be understandable to those who must use them.
7. They can provide recognition of good performance.

However, we must use care in selecting results measurements. Since we are judging performance by some means other than the performance itself, we must be sure that what we are looking at is fairly closely related to performance. With failure measures, this close relationship is quickly lost at lower levels of the organization. Before-the-fact measures tend to retain the close relationship much better.

SAFETY SAMPLING

One of the best methods of fixing accountability, using statistical methods, is safety sampling. Safety sampling measures the effectiveness of the line manager's safety activities, but not in terms of accidents. It measures effectiveness before the fact of the accident by taking a periodic reading of how safely the employees are working.

Like all good acccountability systems or measurement tools, safety sampling is also an excellent motivational tool, for line supervisors find that it is important for their employees to be working as safely as possible when the sample is taken. To accomplish this, they must carry out some safety activities, such as training, supervising, inspecting, and disciplining. Many organizations that have utilized safety sampling report a good improvement in their safety record as a result of the improved interest in safety on the part of line supervisors.

Safety sampling is based on the quality control principle of random sampling inspection, which is widely used by inspection departments to determine quality of production output without making 100 percent inspections. Industry for many years has used this inspection technique, in which a random sampling of a number of objects is carefully inspected to determine the probable quality of the entire production. The degree of accuracy desired dictates the number of random items that will be carefully inspected. The greater the number inspected, the greater the accuracy.

Procedure

1. Prepare a Code The element code list of unsafe practices is the key to safety sampling and supervisor training. This list contains specific unsafe acts which occur in your plant. These are the "accidents about to happen."

The element code list is developed from the accident record of each plant. In addition, possible causes are also listed. The code is then placed on an observation form (see Figure 6-7).

DEPARTMENT

SAMPLING WORKSHEET Page 1 of 1 Safe observations Unsafe acts	D C & Service	Maint. power	Tool room	Foundary & pattern	Stock & shipping	Rotor	Shaft	Punch press	Body & frame	Bracket	Small winding	Large winding	Small assembly	Lg. assem. & pck.
(1) Improper lifting														
(2) Carrying heavy load														
(3) Incorrect gripping														
(4) Lifting w/o protective wear														
(5) Reaching to lift														
(6) Lifting and turning														
(7) Lifting and bending														
(8) Improper grinding														
(9) Improper pouring														
(10) Swinging tool toward body														
(11) Improper eye protection														
(12) Improper foot wear														
(13) Loose clothing--moving parts														
(14) No hair net or cap														
(15) Wearing rings														
(16) Fingers/hands under dies														
(17) Operating equip. at unsafe speeds														
(18) Foot pedal unguarded														
(19) Failure to use guard														
(20) Guard adjusted improperly														
(21) Climbing on machines														
(22) Reaching into machine														
(23) Standing in front of machine														
(24) Leaning on running machines														
(25) Not using push stick (jigs)														
(26) Failure to use hand tools														
(27) Walking under load														
(28) Leaning--suspended load														
(29) Improper use of compressed air														
(30) Carrying by lead wires														
(31) Table too crowded														
(32) Hands and fingers between metal boxes														
(33) Underground power tools														
(34) Grinding on tool rest														
(35) Careless Alum. splash														
(36) One bracket in shaft piling														
(37) Feet under carts or loads														
(38) Pushing carts improperly														
(39) Pulling carts improperly														
(40) Hands or feet outside lift truck														
(41) Loose material under foot														
(42) Improper piling of material														
(43) Unsafe loading of trucks														
(44) Unsafe loading of skids														
(45) Unsafe loading of racks														
(46) Unsafe loading of conveyors														
(47) Using defective equipment														
(48) Using defective tools														
(49) Evidence of horseplay														
(50) Running in area														
(51) Repair moving machines														
(52) No lock-out on machine														
Total unsafe acts Additional unsafe acts														
(53)														
(54)														
(55)														
(56)														
(57)														
(58)														
(59)														
(60)														
(61)														
(62)														

Date________ Time________ Sampler______________

FIGURE 6-7 Safety sampling worksheet

II. Take the Sample The person doing the sampling identifies the department and the supervisor responsible and then proceeds through the area, observing every employee who is engaged in some form of activity and instantaneously recording a safe or an unsafe act. Each employee is observed only long enough to make a determination, and once the observation is recorded, it should not be changed. If the observation indicates that the employee is performing the job safely, he or she is counted on a theater counter. If the employee is observed performing an unsafe act, a check is made on the sampling worksheet indicating the type of unsafe practice by the element code number.

III. Validate the Sample The number of observations required to validate is determined by a preliminary survey and the degree of accuracy desired. The following data must be recorded on the preliminary survey: (1) total observations and (2) unsafe observations. The percentage of unsafe observations is then calculated. Using this percentage P and the desired accuracy, which we will determine as plus or minus 10 percent, we can calculate the number N of observations required by using the following formula:

$$N = \frac{4(1 - P)}{Y^2(P)}$$

where N = total number of observations required
P = percentage of unsafe observations
Y = desired accuracy

(See Figure 6-8 for a table based on this formula.)

For example, if the results produced by the preliminary survey were 126 total observations and 32 unsafe operations, the percentage of unsafe observations would be 32 divided by 126, which is 0.254, or 25 percent. Thus

$$N = \frac{4(1 - P)}{Y^2(P)}$$

$$= \frac{4(1 - 0.25)}{(0.10)^2(0.25)}$$

$$= \frac{3}{0.0025}$$

$$= 1{,}200 \qquad \text{(no. of observations required)}$$

Thus this study must have a minimum of 1,200 observations to give effective results.

IV. Report to Management The results can be presented in many different forms; however, the report should include the following:

Percentage of unsafe observations	*Observations needed*	*Percentage of unsafe observations*	*Observations needed*
10	3,600	30	935
11	3,240	31	890
12	2,930	32	850
13	2,680	33	810
14	2,460	34	775
15	2,270	35	745
16	2,100	36	710
17	1,950	37	680
18	1,830	38	655
19	1,710	39	625
20	1,600	40	600
21	1,510	41	575
22	1,420	42	550
23	1,340	43	530
24	1,270	44	510
25	1,200	45	490
26	1,140	46	470
27	1,080	47	450
28	1,030	48	425
29	980	50	400

FIGURE 6-8 Number of observations needed for 90 percent degree of accuracy

SAFE PRACTICE SAMPLING REPORT

Plant I

Period covered October

Department supervisor	Unsafe practice code number															Observations		Percentage unsafe
	1	2	7	9	11	17	26	34	36	59						Total	Unsafe	
E. Jones - supt.																1,094	39	3.4
Smith - gen.for.																246	9	3.5
Jolas			1					1	1							90	3	3.2
Johnson		3			1			1	1							156	6	3.8
G.McArthur																226	11	4.6
Mantle				1		1		1	1							101	4	3.8
Williams				1	1											53	2	3.6
Nedstrom					1			1	1	2						72	5	6.5
Mack																284	13	4.4
Peters		1			3	1										96	5	5.0
Sadelri							3	1	1							73	5	6.4
Albert	1		1													64	2	3.0
Anderson	1															51	1	1.9

FIGURE 6-9 Safety sampling report

SAFETY SAMPLING REPORT

Plant I

Month of October

Department	Total observations	Unsafe observations	Percentage of unsafe activity	
			This month	Previous month
Manufacturing - Prod	442	77	17.4	12.1
Press	1,815	244	15.3	19.7
Assembly	1,699	59	4.0	4.0
Welding	322	70	21.0	11.2
Subtotals	4,278	450	14.4	14.2
Production Eng.	339	55	16.2	21.5
Plant Engineering	341	51	14.9	26.7
Subtotals	680	106	15.6	23.6
Plant totals				

FIGURE 6-10 Safety sampling report

1. Total percentage of unsafe activity by department and by shift

2. Percentage of unsafe activity by supervisor, general supervisor, or superintendent

3. Number and type of unsafe practices observed

4. Breakdown of types and number of observations of unsafe acts by supervisory responsibility

For examples of these, see Figures 6-9 to 6-12.

Correlation of Results

In 1967 I collaborated with Paul Mueller, corporate safety director of the Green Giant Company, and Jim Young, safety consultant for Employers Insurance of Wausau, on an experiment with this tool of safety sampling. The experiment was cut short and did not accomplish its stated objectives, but it did give us an opportunity to study the tool thoroughly and to correlate our findings with other normal indicators of safety performance. Sufficient samples were taken to ensure validity, and the results were then compared with cost, the all-accident rate, number of accidents, and cost per worker-hour. The results of the experiment are shown below:

Correlation of:	
Accident claim costs with the safety sample taken	+0.353
Accident claim costs with the all-accident rate	+0.364
Number of accidents with the sample taken	+0.446
Accident cost per worker-hour with the sample taken	+0.401
Accident cost per worker-hour with the all-accident rate	+0.245

The results are interesting and seemingly significant. Sampling seems to show the same trends as claim costs, number of accidents, and accident cost per worker-hour, although it correlates better with the all-accident rate (all reported accidents per 1,000 worker-hours).

This seems to mean that sampling provides an excellent indicator of accident problem areas—before the accidents occur. Of course, by far the best value of sampling is motivational. Sampling has been found to arouse extreme interest in safety where there was little interest before.

Sampling also seems to create a great deal of interest among supervisory personnel, for management knows that sampling is perhaps the best indicator of what the supervisor is actually doing about safety. It is a measurement of worker activity, and that is the best measure of supervisory success in safety.

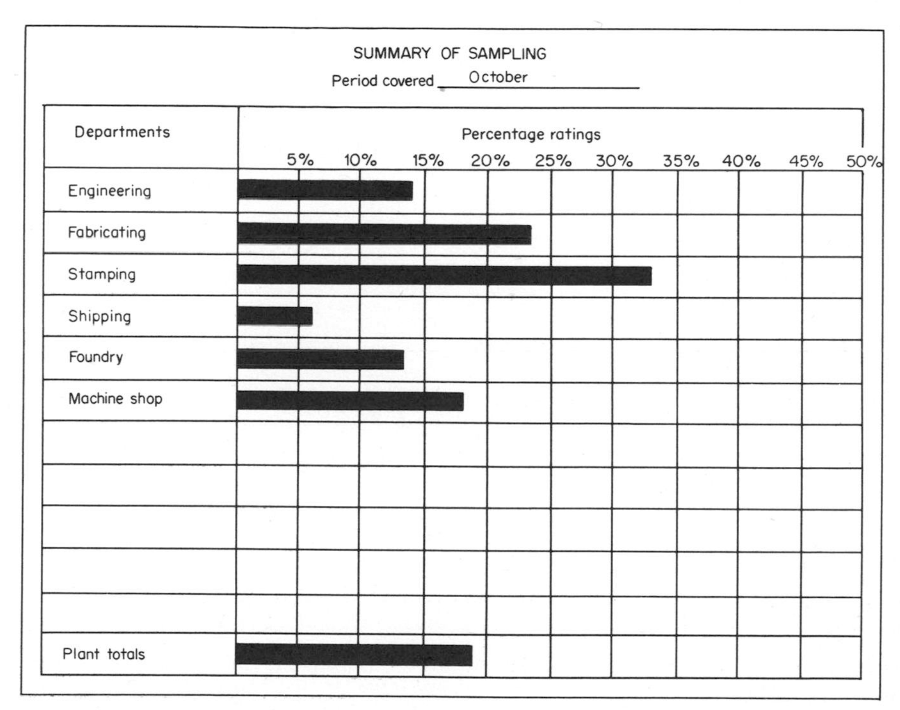

FIGURE 6-11 Safety sampling report

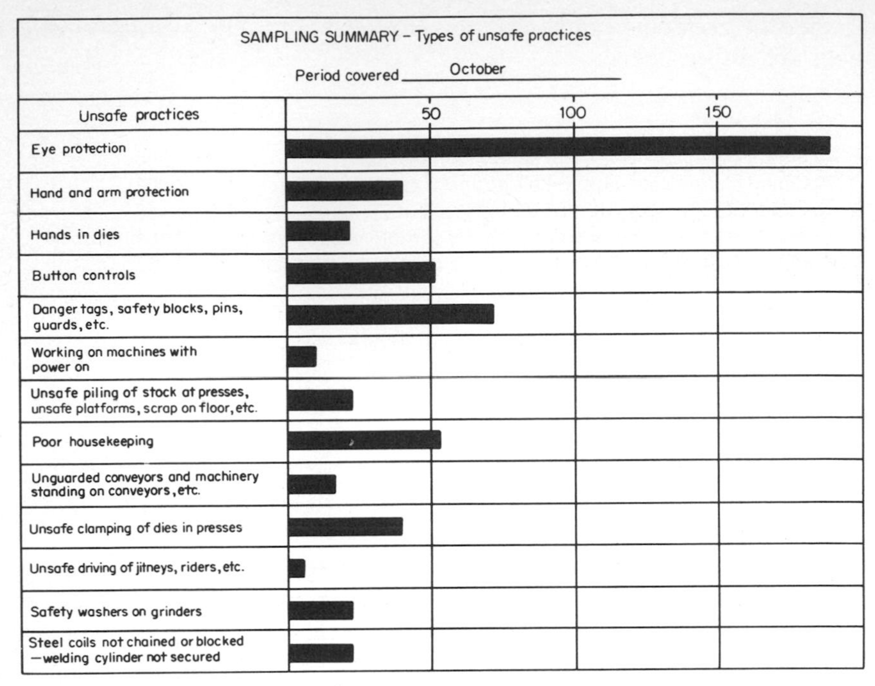

FIGURE 6-12 Safety sampling report

This chapter has only scratched the surface in this important area of measurement. In addition to the techniques touched on here, many more are worthy of study. The sources listed below discuss many excellent new techniques that were not covered in this chapter.

REFERENCES

Allison, W. W.: "High Potential Accident Analysis," *National Safety News,* December 1965.

Bird, F. E.: "Property Damage: Safety's Missing Link," *National Safety News,* September 1966.

Blake, R. P.: *Industrial Safety,* Prentice-Hall, Inc., Englewood Cliffs, N.J., 1943.

Mueller, P. J.: "Supervisor's Incident Investigation Report," Green Giant Company, Le Sueur, Minn., 1968.

Polina, Vincent: "Safety Sampling," *Journal of the ASSE,* August 1962.

Stone, J. R.: "Safety Catalyst," *Journal of the ASSE,* March 1964.

Vilardo, Frank J.: "Some Guidelines to Sampling," *National Safety News,* November 1966.

Weaver, D. A.: "How to Conduct TOR Analysis," Employers Insurance of Wausau, Wausau, Wis., 1967.

7

MEASURING THE MIDDLE AND UPPER MANAGER'S PERFORMANCE

As we shall use it here, the term "middle manager" is a bit of a catchall, meaning an employee at almost any level of management *over* the first-line supervisor and *under* the top executive. Thus a middle manager could be a line manager over two or three supervisors, a plant superintendent with 10 supervisors, or a manufacturing manager or vice president to whom several levels report. Middle managers, then, do not have ultimate policy decision-making authority; rather, they carry out the plans, policies, and procedures decided upon above their level. Similarly, they are not responsible for daily supervision of employees; people under them do that.

How do we measure the middle manager? First, using the same approach discussed in Chapter 6, we can measure performance, results, or both.

As we ascend the organizational ladder, our criteria for measures of safety performance change slightly, and the measures change also. Since the data base grows with each step up, we can begin to use results measures with more confidence. At the lower levels we tend to use performance measures to a large degree to ensure that we are not rewarding only luck. At the top levels we become much more results-oriented and tend to look less at performances. At the middle level we look closely at both performances and results. As top managers judge middle managers, they want to make sure that the middle managers keep a strong safety emphasis (through performance measures), but they probably will also judge them heavily by the results they have achieved.

We use different measures as we go higher up the organizational ladder for several reasons. The results statistics tend to become more reliable and valid at the upper levels. At the lower levels the data base is simply too small to provide statistical significance. Also, at the supervisory level we want to motivate and reward performance, and therefore we must measure performance itself whenever possible. Examples of middle-management measures have been presented in other chapters also:

- In Chapter 4 we discussed, for instance, the prorating of insurance premiums. This is an excellent results (failure) measure that works well at this level, where the use of dollars as a measure is very effective.
- The construction jobsite survey, also discussed in Chapter 4, is an example of a performance-based middle-management (job superintendent) measure.
- Some of the measures discussed in Chapter 6, as well as those discussed in the next chapter, could be used on the middle-management level.

MEASURES USED WITH NEW SAFETY APPROACHES

In recent years a number of new approaches to safety programming have come on the scene, necessitating the use of new measures to make them work and to ensure that supervisory and middle-management functions are in fact being performed. We shall briefly discuss a few of these new approaches.

Safety by Objectives

Safety by objectives (SBO) is one of the most effective of the approaches adopted by organizations in the recent past. It is based on the old management by objectives (MBO) approach. The concept is simple; managers devise specific objectives initially (results, performance, or both), and they and their immediate boss reach a mutual agreement on those objectives. Thus under SBO the measurement problem is simple—it is a part of the whole process. What is measured is simply whether the manager has in fact met the objectives. Only enough record keeping is needed to keep track of progress toward the agreed-upon goal.

Care Control

Care control is a new, somewhat radical approach to safety. It is built on the concept that if employees really believe that management and their immediate supervisors are sincerely interested in them and in their safety, they will perform in a safe fashion. There is a great deal of research from the behavioral sciences to back up this rather novel and simplistic approach to safety. Under a care-control program, supervisors are required, regularly and routinely, to administer positive "strokes" (reinforcements) to each of their people. This can be done in a number of ways, and these are taught to all supervisors in an initial training program, which is based on the theories of transactional analysis. Once supervisors learn the techniques, they are measured to ensure that they do in fact use them (administer the strokes). The measurement is simple. Supervisors are required to record their performance (self-monitoring), and middle managers are measured by the performance of their supervisors. It is a performance measure, the performance being the administering of strokes. When a results measure is used in care control, it is in the nature of an attitude survey of employees. No traditional safety measures are used, except at the corporate level, where traditional measures track progress being made. Care control has resulted in substantial reductions in accidents and costs. In short, it works.

PERFORMANCE MEASURES

While we begin to emphasize results measures more at the middle-management level, some performance measures should be retained here also. Performance measures at this level are simply measures of what middle managers do—of whether they perform the tasks that it is necessary for them to perform. Thus everything we have said about supervisory performance measures applies also at the middle-management level.

However, supervisory performance and middle-management performance should be quite different, necessitating some changes in the measures used at this level. Normally, we want the middle managers to get their subordinates (supervisors) to do something in safety. Thus we can measure them to see whether they meet with their supervisors, check on these supervisors, monitor the quality of the supervisors' work, etc.

AUDITS

A measure of performance (and results) that enters the scene at the upper- and middle-management levels (management of a location, for instance) is the audit of safety performance. One example of such an audit is the Safety Program Appraisal for Multiplant Operations presented in Appendix A.

SAFER

Another example of a safety audit is SAFER, used by Grumman Aerospace Corporation and developed by Robert J. Mills, corporate safety engineer. Following is his description of SAFER:

Evaluation and Audit of the Safety Program

"Safety Audit Factor Evaluation Report" (SAFER)—Safety is essentially a line management function, and local operational managers must accept the responsibility to implement corporate policy and procedures. Local management can be assisted in its task of evaluating the work environment by the Safety Audit Factor Evaluation Report (SAFER), which acts as a guide for the self-evaluation and audit. Within the line organization, supervisors must accomplish predetermined safety objectives and are encouraged to treat safety objectives and production efficiency as inseparable.

The corporate safety manager, working with local managers and safety supervisors, renders a consultative service to each location and provides technical support essential to the success of the local program.

Corporate policy directives prescribe the structure, functions, and procedures for the local safety organization. These broad policy guidelines provide the balanced program which is essential in responding to the requirements of local, state, and federal regulations.

The SAFER audit report is a guide for reviewing and checking key factors in the current loss-prevention activities on a broad category evaluation of "none," "minimal," and "adequate." The evaluation is done annually by the local plant manager and safety supervisor. Items which are evaluated as less than adequate should be incorporated into the safety objectives for the coming year. This exercise enables the local plant safety supervisor to review the status of past objectives, as well as project new items to be worked on to resolve current problems.

The report creates a very definite involvement and commitment action, interrelating the local plant manager and his safety supervisor. Their combined efforts are then stated as an official status report of what they think has to be done, and it enables their superiors at divisional and corporate headquarters to review, concur, or advise on the merits of their conclusions.

Another Plus

Another plus is that the completed safety objectives can also be presented as evidence of "good faith" in regard to complying with the provisions of the OSHA law. The following is a cross-sectional sampling of the eight major sections of the SAFER report, which contains 77 specific items to be rated within the local plant safety program activities. A guide is also furnished to assist local plant management in developing meaningful objectives from the SAFER report. The guide contains more detailed explanations and definitions of the corresponding elements in the SAFER report. Beneath the specific items are questions lettered (a), (b), (c), etc., contained in the guide, which are to be considered before a rating is given to each specific item. [See Figure 7-1.]

This overall plan with the described management control tools and support material, properly administered, has a proved success record with various large corporations. Obviously, it cannot be implemented and bring improved changes overnight. You cannot go from 25 percent loss control to complete control immediately. A corporate safety manager is a quality improvement specialist and must be an integral part of management. The big problem in management today is determining what level of reliability and efficiency is necessary. Any system is a number of parts and subparts working in unison to support the whole. Each function in a management system can be evaluated for excellence. The avoidance of performance error is the key to treating the cause instead of the symptom. Therefore, a corporate safety manager properly handling the described program will not only reduce losses but also improve performance.

Chevron's Review

Another example of the audit of safety performance is the safety program and evaluation used by Chevron Oil Company. It has the following purposes: (1) to determine "soft spots," (2) to gain insights into employee attitudes regarding the effectiveness of the safety program, (3) to give local management direction in making the program more effective, and (4) to assist top management in determining how to improve its support efforts.

The Chevron method is slightly different from some of the other audit approaches. First, a review team is formed consisting of a staff safety engineer and one or two persons responsible for coordinating or implementing the safety program of the organization being reviewed. The next step is to administer a questionnaire to all employees and supervisors (see Figure 7-2).

Next, the review team conducts the program review by means of interviewing key management people, supervisors, and employees on a random basis; attending a safety committee meeting; reviewing all loss-prevention files, accident logs, and reports; observing people at work; tabulating and analyzing the completed survey questionnaire; making inspections of facilities; and making spot reviews of safety suggestions, operating procedures, work orders, and training programs. After this, the safety program review and evaluation is prepared with the assistance of key management people. On the basis of this evaluation, an overall profile of the safety program is constructed, and ratings are determined. Figure 7-3 shows sections A.1 to A.3 of the worksheet (covering policy; employee testing, selection, and placement; and management involvement). The following is an outline of all the areas covered on the worksheet:

A. Organization and administration

   A.1 Statement of policy and responsibilities assigned

   A.2 New and transferred employee testing, selection, and placement

   A.3 Direct management involvement and support

   A.4 Emergency plans

   A.5 Safety rules and standards

   A.6 Loss-prevention organization

B. Hazard control

   B.1 Housekeeping

   B.2 Guarding

   B.3 Physical work-area protection

   B.4 Material handling

   B.5 Personal protective equipment

   B.6 Fire protection

   B.7 Environmental health

The major sections of the Safety Audit Factor Evaluation Report (SAFER) are:

		None	*Minimal*	*Adequate*

A. *ORGANIZATION* (five items listed)

Ex. Item: #3. Safety committees effectively organized at all levels . . 3 ☐ ☐ ☐

(a) Are all levels of management involved in regularly scheduled safety meetings?

(b) Do first-line supervisors conduct safety meetings with their employees routinely?

(c) Is relevant safety information exchanged between employees and management?

B *ADMINISTRATION* (18 items listed)

Ex. Item: #1. Local management actively participates in safety effort 1 ☐ ☐ ☐

(a) Plant/facility manager or representative are part of management safety committee?

(b) Plant/facility manager reviews all significant accidents?

(c) Plant/facility manager actively supports efforts of supervisors on safety matters?

C. *CONTROL OF PHYSICAL HAZARDS* (17 items listed)

Ex. Item: #3. Approved type personal protective equipment available and used . . . . 3 ☐ ☐ ☐

(a) Equipment conforms to established standard: ex. Respiratory protection has Bureau of Mines approval; hard hats meet ANSI Z89.1; eye protection meets ANSI Z87.1; safety footwear meets ANSI Z41.1

(b) Areas requiring the use of protective equipment are properly posted?

(c) Personal protective equipment is routinely examined for defects and indications of deterioration?

(d) Employees are fully knowledgeable as to why equipment is required?

D. *CONTROL OF OCCUPATIONAL ENVIRONMENTAL HAZARDS* (13 items listed)

Ex. Item: #1. Atmospheric surveys conducted periodically (i.e., air, dust, gases) . . . . 1 ☐ ☐ ☐

(a) Plant monitored for concentration of materials identified in OSHA Standard 1910.93 and accurate records of results maintained?

(b) Where measurements show concentrations in excess of established limits, engineering controls initiated to reduce the level?

(c) Records of environmental tests and samples available for review? (Records maintained for 20 years)

E. *INVOLVEMENT AND DEVELOPMENT* (10 items listed)

Ex. Item: #4. Safety indoctrination given to all new or transferred employees . . . . 4 ☐ ☐ ☐

(a) Each new employee given a copy of the general plant/facility safety rules?

(b) Job safety analysis reviewed with supervisor?

FIGURE 7-1 SAFER outline

(c) All relevant departmental operating procedures reviewed with supervisor?

(d) Employee given apprentice status until he demonstrates capability to accomplish job assignment unassisted?

Ex. Item: #7. First aid techniques taught to all employees .... 7 □ □ □

(a) Selected individuals given thorough first aid training and respond to plant/facility emergency?

(b) All employees instructed in techniques of emergency life saving?

F. *MOTIVATION* (10 items listed)

Ex. Item: #1. Posting of safety information adequate in volume and strategically located .... 1 □ □ □

(a) Information as required by OSHA posted in a place accessible to all employees?

(b) Accident prevention signs posted as required by OSHA Standard 1910.145?

(c) Excerpts from general safety committee meeting minutes posted for employees?

G. *ACCIDENT EVALUATION AND REPORTING* (11 items listed)

Ex. Item: #4. Plant/facility management carries on in-depth analysis for reasons of incidents/ accidents .... 4 □ □ □

(a) Medical data evaluated and correlated to ascertain any pattern to health problems?

(b) Statistical analysis conducted on factors related to accidents to determine and pattern?

(c) Production systems analyzed to eliminate hazards?

(d) Employee work schedule evaluated for causal factors?

Ex. Item #7. Records maintained of all accidents and illnesses .... 7 □ □ □

(a) Maintain records per OSHA (29 CFR 1904) for 5 years.

(b) Notation made in employee medical records (certain records must be kept for 20 years).

(c) File of accident investigation reports maintained for statistical evaluation?

NOTE: Each major section has space for local management to include any additional items they feel are necessary or appropriate.

Projected Safety Objectives: After the Safety Audit Factor Evaluation Report (SAFER) is completed, review all items rated "None" and "Minimal" and decide which specific items need the most attention. A list of these items constitute the safety objectives that need priority attention and should be stated in terms of specific programs or procedures to be implemented. For example:

SAFER Ref: No.		Priority	Project Completion Date
	C—Item #4. Develop and institute Hazard Operation Permit System for all hazardous operations. System to be coupled with existing work order system.	1	September 1

Safety Survey

How good is our safety program? Here is a chance to express your personal opinion on how you feel toward safety.

Finish each sentence to show what you think of the program. You may want to give examples to explain your answers.

There is no need to sign this survey. Your answers will be combined with those of others to determine the group's opinion of our safety program.

Be frank!

Tell us where our program has failed; tell us our good points, too! How can we do a better job in safety?

FIGURE 7-2 Chevron's employee questionnaire

B.8 General chemical hazards

B.9 Hazard identification and analysis

B.10 Safe work permits and confined space entry

B.11 Equipment safety devices

B.12 Maintenance

B.13 Solid waste, air and water pollution, and spill control

B.14 Design safety (new and modification projects)

C. Training and motivation

C.1 Indoctrination of new and transferred employees

C.2 Employee training

C.3 Supervisor safety training

C.4 Safe operating procedures

C.5 Internal self-inspection or auditing

C.6 Safety meetings

C.7 Employee-supervisor safety contact and communication

C.8 Safety suggestions

C.9 Safety recognition and promotion

1. Safety training in my department...

2. My knowledge of accident prevention...

3. My boss's views on safety...

4. The best part of our safety program is...

5. Cooperativeness of my work group in achieving safety...

6. Evidence of safety in my work area...

7. Rewards for good safety performance...

8. Management's concern for safety...

9. A person involved in a series of accidents...

10. We hear about safety...

11. My pet peeve about the safety program...

12. I think we can improve our safety program by...

FIGURE 7-2 Chevron's employee questionnaire (*continued*)

D. Accident investigation and cause analysis

D.1 Accident investigation by supervisor

D.2 Accident cause analysis

D.3 Claims investigation and follow-up

D.4 Reporting and record keeping

PROGRAM REVIEW WORKSHEET

Company ______________ **Organization** ________________ **Location** __________

Date of review _________ **Reviewed by** ______________________________________

A. ORGANIZATION AND ADMINISTRATION	NOTES/COMMENTS
A.1 *Statement of Policy and Assigned Responsibilities* ___ Written ___ Communicated and understood ___ Posted and current ___ Supporting directives Management and Supervisor Responsibility ___ Assigned and defined ___ Performance planning review ___ Accountability	
A.2 *New and Transferred Employee Testing, Selection, and Placement* ___ Preemployment physicals ___ Job-related skills tests ___ Review of past safety and driving record; i.e., discuss with previous employers, check for D.M.V. violations, etc. ___ Base-level audiograms, blood cholinesterase norm. ___ Special medical tests, e.g., EKG, chest x-rays, pulmonary function, and lead urinalysis	
A.3 *Direct Management Involvement and Support* ___ Safety programs/committees ___ Establishment of goals ___ Survey of facilities and work practices ___ Review and analysis of programs ___ Investigation of accidents and review of reports ___ Accountability for safety performance ___ Policy endorsement	

FIGURE 7-3 Chevron's worksheet

E. Off-the-job safety

E.1 Organization and administration

E.2 Investigation, reporting and cause analysis

Figure 7-4 shows the ratings for sections A.1 and A.2. Each of the sections outlined above is given a similar rating. The ratings are then quantified, using preassigned numbers, as shown in Figure 7-5. The numbers can be changed, depending upon what weightings each area is given. In this way, different items can be emphasized in different periods.

The review is followed by a verbal feedback session with the responsible manager and staff. Then a written report is prepared, covering the points made in the closing conference. The report includes a transmittal letter, the safety program review and evaluation forms, and any suggestions for improving the overall effectiveness of the program. A copy of the written report is sent to the appropriate executives and managers.

Safety Performance Indicator

Another approach to measurement that incorporates both results and performances is the Safety Performance Indicator (SPI). It was developed and used by C. D. Attaway, chief safety engineer for Thiokol Chemical Corporation. Following is his description of the SPI:

> The Safety Performance Indicator does not replace the standard disabling injury frequency or severity rate, but is rather a means of combining it with other pertinent and important elements in a systematic fashion to reveal why accidents occur and what management has done to control these causes. This method is geared primarily to factory, plant, or installations of a large operating or manufacturing organization and is designed to highlight those factors that control a corporation accident experience.
>
> An added feature of the formula is the flexibility offered by adding or removing elements. For "within the division" purposes such things as promptness or completeness of accident reporting can be made an element. Or violations of safety rules and regulations and other circumstances that impede progress can be included as factors in the formula.
>
> The intent of the method is simply to indicate whether the performance of the safety program at an activity or operation is getting better or worse and *why*.

Explanation and Computation of the Formula

The Performance Indicator is expressed by the formula $PI = A + DL + PD + RC + MD + S$. All data used in this formula can be obtained from accident reports. These reports are evaluated as submitted, with nothing added or deleted. Accidents and their consequences which are beyond a manager's immediate resources are not included.

SAFETY PROGRAM REVIEW AND EVALUATION

Company______ **Organization**________ **Location**_________

Date of review ________ **Reviewed by** ________________________

A. ORGANIZATION AND ADMINISTRATION	COMMENTS
A.1 *Statement of Policy and Assigned Responsibilities* ____ POOR: No statement of safety and health policy. Responsibility and accountability not assigned. ____ FAIR: A general understanding of responsibilities and accountability, but not written. ____ GOOD: Policy and responsibilities written and distributed to all employees. Safety record is considered in employee performance evaluations. ____ EXCELLENT: Safety and health policy is reviewed annually, posted, and understood by employees. Responsibility and accountability are emphasized in employee performance evaluations and salary treatment.	
A.2 *New and Transferred Employee Testing, Selection, and Placement* ____ POOR: Only preemployment physical examinations are given. ____ FAIR: Job-related skills tests are administered when applicable. ____ GOOD: Past safety record considered by discussing with previous employers when possible and checking D.M.V. violations when applicable. Medical tests such as audiogram and blood cholinesterase given to establish base levels when appropriate. ____ EXCELLENT: Safety attitude and record are considered. Special medical tests such as EKG, chest x-rays, pulmonary function, and lead urinalysis are given when appropriate.	

FIGURE 7-4 Chevron's rating process

A. The first element in the formula is *A*, which denotes the total number of reportable accidents occurring on work or operations under the jurisdiction of the manager that are within the resources of the corporation to correct. (One day or more actually lost or charged by ANSI Z16.1, or accident resulting in $100 or more property damage.)

B. *DL* denotes the total number of days lost, including fixed charges for permanent impairments and fatalities, as the result of injuries sustained by *on-duty* personnel.

C. *PD* denotes the amount of property damage in dollars to property and/or equipment.

D. *RC* denotes reason for cause, which is each instance where a supervisory and/or physical deficiency produced the accident cause. Physical deficiencies are defined as those accident factors which stem from poor construction practice, maintenance, layout, or equipment. Supervisory deficiencies are defined as those accident factors which stem from failure of supervisors at the job level to direct, inspect, instruct, and place the working force in a manner conducive to safety operation.

These definitions make it obvious that one, two, or none may be assigned a single accident. Selection depends on which factors, if any, are predominant. For example, if a man falls from a scaffold not provided with a guardrail, a physical deficiency is assigned. If a man falls from a scafford while using the cross members as means of access when adequate access has been provided by means of suitable ladders or stairways, a supervisory deficiency is assigned. If a piece of equipment is wrecked because of brake failure while being operated at excessive speed, both a physical and a supervisory deficiency are assigned.

E. *MD* denotes management deficiency, each instance where the report indicates, on the part of management, an apathetic review, a poor grasp of the facts involved, and inadequate directions to correct deficiencies. A management deficiency is also assigned to a report involving a fatality, permanent impairment, or property damage in excess of $10,000 where such incidents occur as a result of a repetitive cause in the plant in a calendar year. For example, if a man loses a finger operating a power saw without a guard and afterward another man loses a finger by the same cause, a management deficiency is assigned.

If a fire occurs because of an unattended salamander stove, with resulting damage in excess of $10,000, and then another fire occurs, with cause the same and costing over $10,000 loss, a management deficiency is assigned. This is a weighty factor in the formula because it is a determined evaluation of report data in terms of application of policy and instructions issued by the general manager to top supervision. Top-management support is the hub around which the division safety program revolves, and therefore it becomes a principal consideration in evaluating safety performance. When reports are evaluated, a management deficiency is not assigned when a manager corrects a deficiency.

F. *S* denotes surcharges, assigned in each instance in which an accident resulted in a fatal, permanent impairment and/or damage in excess of $10,000. Sur-

SAFETY PROGRAM RATING FORM

A. ORGANIZATION AND ADMINISTRATION	Poor	Fair	Good	Excellent	
1. Statement of policy and assigned responsibilities	0	8	14	20	
2. New and transferred employee selection, testing, and placement	0	4	7	10	
3. Direct management involvement and support	0	12	21	30	
4. Emergency plans and security	0	4	7	10	
5. Safety rules and standards	0	4	7	10	
6. Safety organization	0	8	14	20	
Total value of circled numbers	___ +	___ +	___ +	___	= ___ X .30 rating
B. HAZARD CONTROL					
1. Housekeeping	0	4	8	11	
2. Guarding	0	2	3	5	
3. Physical work area protection and plant layout	0	3	6	8	
4. Materials handling	0	2	3	5	
5. Personal protective equipment	0	3	5	7	
6. Fire protection	0	4	8	11	
7. Environmental health	0	4	8	11	
8. General chemical hazards	0	2	3	5	
9. Hazard identification and analysis	0	2	3	5	
10. Safe work permits and confined space entry	0	3	6	8	
11. Equipment safety devices	0	3	5	7	
12. Maintenance	0	2	3	5	
13. Solid waste, air and water pollution, and spill control	0	2	3	5	
14. Design safety	0	3	5	7	
Total value of circled numbers	___ +	___ +	___ +	___	= ___ X .35 rating
C. TRAINING AND MOTIVATION					
1. Indoctrination of new and transferred employees	0	4	7	10	
2. Employee training	0	6	11	16	
3. Supervisor safety training	0	8	14	18	

FIGURE 7-5 Quantification

4. Safe operating procedures	0	6	10	15
5. Internal self-inspection or auditing	0	2	4	6
6. Safety meetings	0	2	4	6
7. Employee/supervisor safety contact and communication	0	6	10	15
8. Safety suggestions	0	4	6	8
9. Safety recognition and promotion	0	2	4	6

Total value of circled numbers ____ + ____ + ____ + ____ = ________ X .20 rating

D. ACCIDENT INVESTIGATION AND CAUSE ANALYSIS

1. Accident investigation by supervisor	0	14	25	35
2. Accident cause analysis	0	12	21	30
3. Claims investigation and follow-up	0	6	10	15
4. Reporting and record keeping	0	8	14	20

Total value of circled numbers ____ + ____ + ____ + ____ = ________ X .10 rating

E. OFF-THE-JOB-SAFETY

1. Organization and administration	0	24	42	60
2. Investigation, reporting, and cause analysis	0	16	28	40

Total value of circled numbers ____ + ____ + ____ + ____ = ________ X .05 rating

Summary

	Maximum
A. ORGANIZATION AND ADMINISTRATION	(30%) ________
B. HAZARD CONTROL	(35%) ________
C. TRAINING AND MOTIVATION	(20%) ________
E. ACCIDENT INVESTIGATION AND CAUSE ANALYSIS	(10%) ________
F. OFF-THE-JOB SAFETY	(5%) ________
	TOTAL RATING ________

PROGRAM MEASUREMENT:

Below 40—Poor or ineffective program

40–70 —Fair program with inconsistent results

70–90 —Good program, results show improving safety performance

90–100—Outstanding safety program with excellent results

charges are also a weighty factor because such serious incidents must be effectively controlled before the safety program can be considered to be making progress. It is conceded that in some instances severity of an injury or amount of property damage is a matter of chance. However, the experience of the corporation establishes that most of our fatalities, permanent impairments, and large dollar losses are foregone conclusions if certain incidents occur.

For example, when a man falls from a considerable height, the severity of his injuries is extreme. The same holds true if he is in contact with equipment which engages a power line. When valuable property is stored without segregation of combustibles or provision of fire protection or where high explosives are processed and explosions occur, the losses are usually large. Since the severity of incidents of this nature is largely predictable and since practical controls are well publicized, the occurrence of such accidents must have special consideration when measuring performances.

G. The computations made from the formula $A + DL + PD + RC + MD + S$ are added together for current performance and compared to a normal. The normal is computed by applying the same formula to accident reports received for three preceding years. After the accumulation of five years' data, the normal will be a five-year rolling average, i.e., the most recent five-year experience. This will tell whether the performance of the installation is better or worse, how it is contributing to the overall performance, and the reasons why the performance is better or worse.

RESULTS MEASUREMENTS

At the middle-management level we concentrate on results measures. Here again we can work with before-the-fact measures or with failure measures. The validity and reliability of our failure measures may depend upon the size of the unit we are working with. Before-the-fact results measures used to judge middle managers' performance would include safety sampling and inspections by staff safety. Failure measures can be used a bit more extensively at this level. Here our traditional indicators become more useful. A number of these are listed in Chapter 8. Also, dollar measures become very useful here as discussed earlier.

PAST PERFORMANCE AS A STANDARD

The past performance of any group is the best standard to use as a guide for its present performance. The people in the group understand it and will accept it, whereas they are often reluctant to accept an arbitrary standard set by someone outside the group. Often they believe that competing against a different group is unfair, since no two groups face the same challenges, hazards, or situations. Management is interested principally in its own company and whether it is improving.

When past performance is used as a standard, it is often difficult to know whether the changes that have occurred are truly significant, reflecting a different supervisory performance, or whether they are merely chance happenings—random fluctuations. There is a technique now in use which helps us to determine this.

Safe-T-Score

The Safe-T-Score is based on a statistical quality control test which is used to examine the means of two groups of comparable data for significant differences. In statistics it is usually referred to as the "*t* test." Here is its formula:

$$\text{Safe-T-Score} = \frac{\text{frequency rate now} - \text{frequency rate past}}{\sqrt{\dfrac{\text{frequency rate past}}{\text{million worker-hours now}}}}$$

The Safe-T-Score is a dimensionless number. A positive Safe-T-Score indicates a worsening record; a negative Safe-T-Score indicates an improved record over the past. Here is how the Safe-T-Score can be interpreted:

- If the Safe-T-Score is between +2.00 and −2.00, we know that the change is not significantly different. The variation is due to random fluctuation only.
- If the Safe-T-Score is over +2.00, we know that the record is significantly worse than it was in the previous period. Something wrong has happened.
- If the Safe-T-Score is under −2.00, we know that the record is significantly better than it was in the last period. Something has changed for the better.

For example, if we compare the all-accident frequency rates of two locations of a company, each against its past record, we start with this information:

Location X	*Location Y*
Last year −10 accidents 10,000 worker-hours	Last year −1,000 accidents 1,000,000 worker-hours
Frequence rate = 1,000	Frequency rate = 1,000
This year −15 accidents 10,000 worker-hours	This year −1,100 accidents 1,000,000 worker-hours
Frequency rate = 1,500	Frequency rate = 1,100

The frequency rate for location X has increased by 50 percent; that for location Y has increased by only 10 percent. Are these figures significant? Has something gone wrong at one or both locations? We do not know—yet. By using the Safe-T-Score, however, we can answer these questions.

$$\text{Safe-T-Score} = \frac{\text{frequency rate now} - \text{frequency rate past}}{\sqrt{\dfrac{\text{frequency rate past}}{\text{million worker-hours now}}}}$$

Location X

$$\text{Safe-T-Score} = \frac{1{,}500 - 1{,}000}{\sqrt{\dfrac{1{,}000}{0.01}}}$$

$$\text{Safe-T-Score} = \frac{500}{317}$$

Safe-T-Score = +1.58

In this case the 50 percent increase is not significant.

Location Y

$$\text{Safe-T-Score} = \frac{1{,}100 - 1{,}000}{\sqrt{\dfrac{1{,}000}{1.00}}}$$

$$\text{Safe-T-Score} = \frac{100}{31.7}$$

Safe-T-Score = +3.17

In this case the 10 percent increase is significant. Something is wrong at location Y.

The Safe-T-Score is a very useful tool in our record system. It allows us to interpret our records much better.

STATISTICAL CONTROL TECHNIQUES

As mentioned previously, the accident rate may fluctuate from period to period and still reflect nothing more than chance variation. There are times, however, when a fluctuation occurs because something is different in the system. This could be a period of high turnover, and hence untrained workers, or of sudden production pressure, for example.

Safety professionals need a tool which will enable them to detect the presence of new accident causes. They should be relatively unconcerned with minor fluctuations in the accident picture when they are sure that the situation has remained stable. However, they need something which will alert them quickly when the situation has become unstable.

Statistical control techniques can perform this essential job. They will signal a significant change in the accident process, giving the safety professional assurance that a change has taken place. The safety specialist can then identify

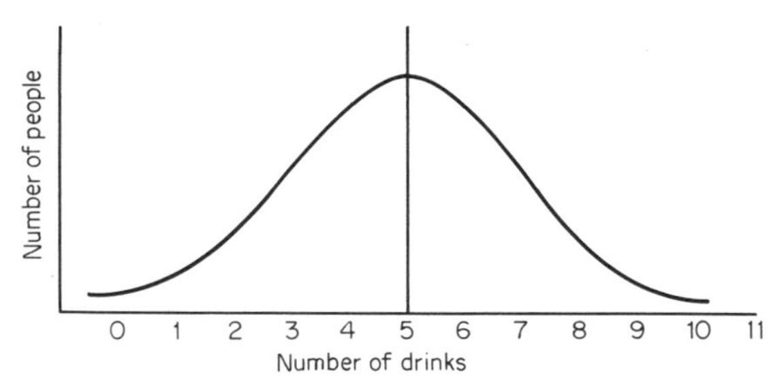

FIGURE 7-6 Normal distribution curve

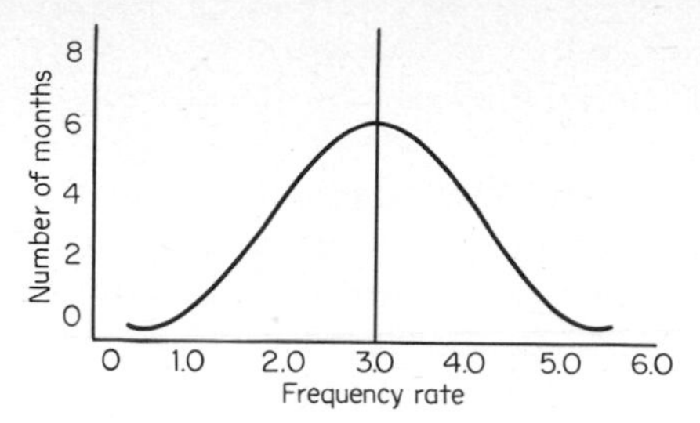

FIGURE 7-7 Normal distribution curve

causes. These techniques have proved themselves in quality control. They can be used equally well in accident control.

Statistical methods can help us to do five things: (1) plan programs for obtaining data so that reliable conclusions can be drawn from them, (2) organize and analyze our raw data to bring out the maximum information, (3) establish or pinpoint cause-and-effect relationships, (4) assess the reliability of our conclusions, and (5) monitor trends and processes.

Statistical Control Charts

In monitoring, the basic tool is the statistical control chart. The control chart is based upon the *bell-shaped curve,* which is applicable to many things, such as height and weight of people, frequency of accidents, or number of drinks consumed in a given period of time.

As an example, let us take the number of drinks consumed by a group at a lengthy cocktail party (see Figure 7-6). The average, let us say, was five drinks per person. A few people had only one drink each (pretty rare), a few people had as many as ten drinks (also, we hope, rare), and most people had between three and seven drinks.

Now we apply the chart in Figure 7-6 to the accident frequency rate of our plant (see Figure 7-7). The average in this case is a frequency rate of 3.0.

We want to put limits on our chart, limits that we decide we should stay within in the future—that is, an *upper control limit* and a *lower control limit* (Figure 7-8). We choose those limits such that if any month's record falls outside them, we know that it is not strictly because of chance—something is wrong.

Suppose we set our limits so that there is only a 1 percent possibility that a month's record will lie outside those limits by chance; we arrive at the shaded area in Figure 7-8. We then turn our curve on its side and plot our months along the bottom, and we have a control chart (Figure 7-9). Each period (biweekly) is plotted. Any time our record goes outside the limits, we know that it is not because of chance—something is wrong—and a red flag goes up.

The control chart is the working tool of statistical control. On it the observed accident rates are plotted against time, with the overall accident rate or mean for the entire period. Finally, upper and lower control limits are computed such that the probability that an accident rate will exceed the limits by chance alone is very small.

From a study of statistical control chart data we can tell whether the system is a relatively constant one. If it is, we have a stable situation. Conversely, we can tell whether the system has changed. If an accident rate exceeds the upper limit,

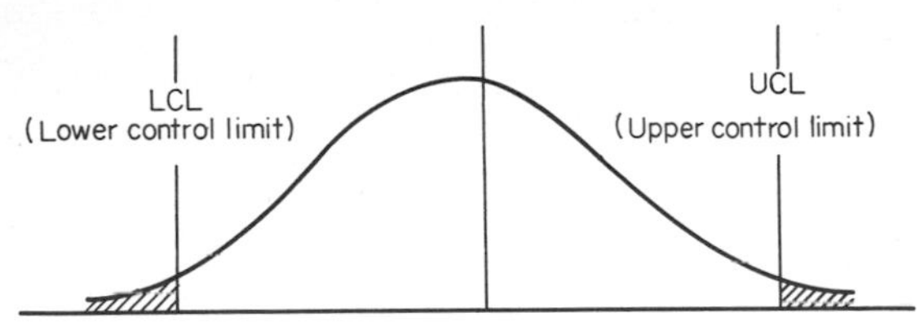

FIGURE 7-8 Control limits

this signals a change for which there is an assignable cause. Similarly, when a point falls below the lower limit, we infer that there has been a significant change for the better.

Statistical control techniques offer a means for making the work of accident reduction more effective and efficient. They cannot assign cause, but they can point out where and when to look for causes.

Preparation of Control Charts The basic data needed for an accident control chart are the number of accidents and the number of hours worked for successive periods. If these data are not available from company records, they will have to be collected on a weekly, biweekly, or monthly basis.

The ratio of accidents to worker-hours for any given period will give the rate for that period. Customarily, 20 such rates should be available before a control chart is drawn up, although there are occasions when a smaller number are used. When these rates, together with the mean accident rate for the total period and upper and lower control limits, are plotted, the resultant chart might resemble the one shown in Figure 7-10.

One percent control limits were used in this instance. This gives a 0.005 probability of the accident rate's falling above the upper limit and a like probability of its falling below the lower limit when the accident rate is stable. If the true accident rate or mean is identical with the estimated rate, the probability of observing a point out of control is 1 percent. (The greater the number of worker-hours covered, that is, the greater the length of time over which data are collected, the more closely the estimated rate will approximate the true rate.) Under these conditions only 1 rate in 100 will be out of control. Should the underlying accident rate change, the probability of observing a point out of control is more than 1 percent.

One percent control limits have proved satisfactory for industrial accident studies, but should a situation arise in which all data points are found to fall within control limits, 5 percent limits might prove more satisfactory. Smaller differences in accident rates between periods would be detected, but at the price of more false signals, since the probability of a point's being out of control when there is no change is 5 percent versus 1 percent for the 1 percent limits.

Control Limits Both 1 and 5 percent limits can be computed very easily with a form similar to the one shown in Figure 7-11. The following computations should be carried out:

For 1 percent limits:

1. Enter in column 1 the number of accidents (X) occurring in each period.
2. Enter in column 2 the number of worker-hours (MH) in each period.
3. Divide the sum total of column 1 by the sum total of column 2. This will give

$$A = \frac{X}{\text{MH}}$$

4. Subtract A from 1 and multiply by A:

$$A(1 - A)$$

5. Divide the product of step 4 by the number of worker-hours for each period and enter in column 3.

$$\frac{A(1 - A)}{\text{MH}}$$

6. Find the square root of each entry in column 3 and enter in column 4.

$$\sqrt{\frac{A(1 - A)}{\text{MH}}}$$

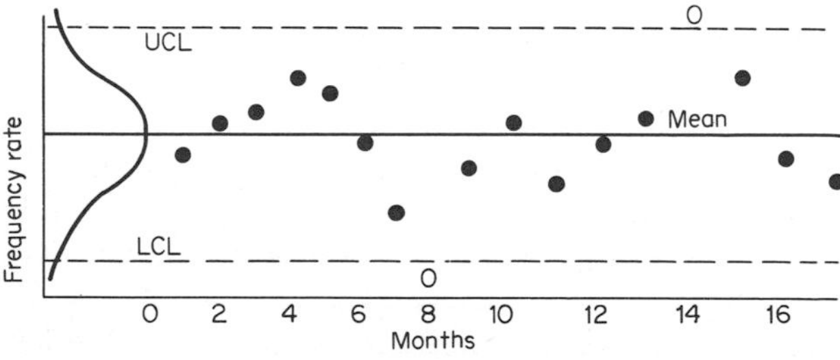

FIGURE 7-9 Control chart

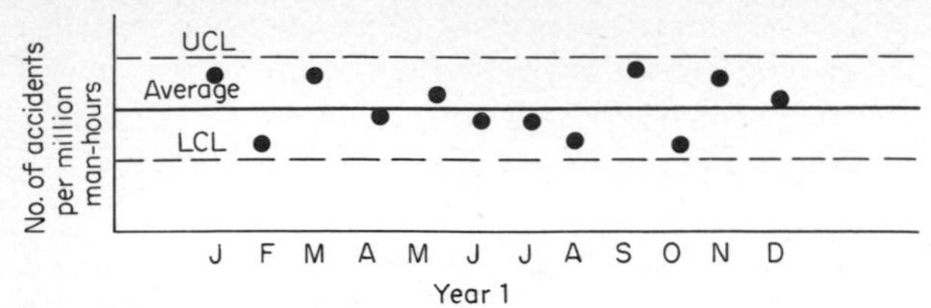

FIGURE 7-10 Control chart for accident frequency, XYZ Manufacturing Company

7. Multiply each entry in column 4 by 2.576 and enter in column 5.

$$2.576\sqrt{\frac{A(1-A)}{\mathrm{MH}}}$$

The upper control limit for each period is A plus the corresponding entry in column 5.

$$\mathrm{UCL} = A + 2.576\sqrt{\frac{A(1-A)}{\mathrm{MH}}}$$

The lower control limit for each period is A minus the corresponding entry in column 5.

$$\mathrm{LCL} = A - 2.576\sqrt{\frac{A(1-A)}{\mathrm{MH}}}$$

For 5 percent limits, substitute 1.96 for 2.576 in step 7. Procedures for computing are otherwise identical.

It takes little more than a glance at year 1 of Figure 7-12 to see that the

Time interval	(1) Number of accidents	(2) Man hours worked	(3) $\frac{A(1-A)}{MH}$	(4) $\sqrt{\frac{A(1-A)}{MH}}$	(5) $2.576\sqrt{\frac{A(1-A)}{MH}}$	(6) UCL	(7) LCL
(1)							
(2)							
(3)							
(4)							
(5)							
(6)							
(7)							
(8)							
(9)							
(10)							
(11)							
(12)							
Total							

FIGURE 7-11 Sample form for calculating control limits

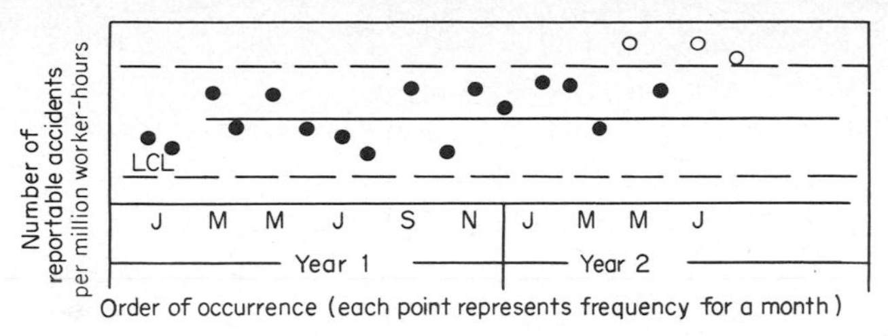

FIGURE 7-12 Control chart for accident frequency

accident rate was very stable throughout the year. No rate falls outside the limits. At this moment in time the safety director would have had every assurance of a well-controlled accident situation but would have had no knowledge of what the future might hold.

Let us suppose that the safety director continued to use this chart and plotted data points as the months passed. By the middle of year 2 the chart would have taken on quite a different appearance (Figure 7-12). The rate of April of year 2 can be seen to signal a change. This was the time to examine the situation closely. In this particular instance the records disclosed that a plant expansion program had been instituted for which no provision had been made in the safety program.

Once a rate has fallen outside the limits, signaling a change to the chart user, there is no way of telling at that time when stability will be reestablished and at what level. If subsequent points fall within the limits, this indicates one or two possibilities: (1) either the cause of the change was determined and corrected, or (2) the cause was transitory and self-correcting. Should subsequent points continue to fall outside the limits, one might infer that the reestablishment of stability has taken place at another level, but this could not be concluded until several points have accumulated. For this reason, practice calls for delaying the computation of a new mean and new control limits until at least 10 such points have accumulated.

To distinguish between instability and a new level of stability, one can apply control limits. If all the points falling outside the original limits are contained by these new limits, one can assume a new level of stability, but if they are not so contained, there is continuing instability.

REFERENCES

Attaway, C. D.: "Safety Performance Indicator Fills a Management Need," *Journal of the ASSE,* March 1969.

Dunlap, J. W.: *Manual for the Application of Statistical Techniques for Use in Accident Control,* Dunlap & Associates, Stamford, Conn., 1958.

Martin, J. A.: "Large Plant Safety Program Management," *Journal of the ASSE,* May 1963.

Mills, Robert: "Setting up and Auditing a Corporate Safety Program," *Journal of the ASSE,* October 1973.

Odiorne, George: *Management by Objectives,* Pitman Publishing Corporation, New York, 1965.

Petersen, Dan: "Safety by Objectives (SBO)," in *Safety Management: A Human Approach,* Aloray Publisher, Englewood, N.J., 1975.

Recht, J. L.: "Systems Safety Analysis," *National Safety News,* April 1966.

Rockwell, T. H.: "Safety Performance Measurement," *Journal of Industrial Engineering,* January 1959.

———: "A Systems Approach to Maximizing Safety Effectiveness," *Journal of the ASSE,* December 1961.

Schowalter, E. J.: "A Year's Trial with a New Safety Measurement Plan," *Journal of Occupational Medicine,* June 1966.

Smart, D. O.: "Care Control," *Occupational Hazards,* June 1976.

Standard Oil Company: *Safety Program Review and Evaluator Guide,* 1976.

Tarrants, William: "Application of Inferential Statistics for the Appraisal of Safety Performance," *Journal of the ASSE, March 1967.*

———: "Applying Measurement Concepts to the Appraisal of Safety Performance," *Journal of the ASSE,* May 1965.

8

MEASURING COMPANYWIDE SAFETY PERFORMANCE

Measures of companywide safety performance can give us information concerning the company's internal performance, or they can allow us to compare the performance of one company with that of other organizations. We shall look at both these aspects of measurement in this chapter. First, we shall consider traditional measures of companywide safety performance.

FREQUENCY AND SEVERITY RATES

Traditionally, we have used two figures to measure companywide safety performance: the frequency rate, which is the number of disabling injuries (lost-time cases) per million worker-hours, and the severity rate, which is the number of days lost, or charged, per million worker-hours. For years these have been governed by the American National Standards Institute Standard Z16.1 and more recently by slightly changed rules under OSHA. The National Safety Council has also judged our national progress with these measures for some time. Each year, in "Accident Facts," it publishes the national frequency and severity rates, thus giving us a picture of the progress we have made in safety in the country as a whole.

As indicators of internal performance, these rates have some serious weaknesses. The validity and reliability of these measures are a function of the size of the company, or how large the data base is, but the real weakness seems to be the fact that the rates are not meaningful to people in the organization. Those in top management often do not understand the rates and may wonder why their safety people cannot talk like managers and use more meaningful measures. As indicators for use in comparing the company's progress with that of other organizations, the major difficulty has to do with input integrity. Each organization seems to compile its safety records according to its own rules, regardless of the details of the Z16.1 standard. It is common knowledge that when we attempt to compare our progress with that of other organizations using these measures, we have no idea how the other organizations have kept their records. Many organizations do not keep their records according to the Z16.1 standard and have no intention of doing so. Still, frequency and severity rates are the final measures used in most record systems today. Since these measures are very weak, however, we ought to examine other available measures, such as the following:

1. *Frequency-severity indicator (FSI).* This is a combined frequency and severity rate. FSI equals the square root of the frequency rate times the severity rate divided by 1,000:

$$\text{FSI} = \sqrt{\frac{F \times S}{1{,}000}}$$

For example:

Frequency rate	*Severity rate*	*FSI*
2	125	0.5
4	250	1.0
8	500	2.0

2. *Total cost of first-aid cases.* In this appraisal the pro rata cost per case handled in-plant would be considered. For example, say that a first-aid unit in a plant costs $10,000 to operate. Forty percent of the unit's time is used to treat first-aid cases. An average of 1,000 cases are treated; therefore, the cost is $4 per case. This example combines industrial and nonindustrial first-aid cases. If average time per case is substantially different for each of these categories, a separate average cost per case may be better.

3. *Cost incurred.* This includes the actual compensation and medical costs paid for cases which occurred in a specified period plus an estimate of what is still to be paid for those cases.

4. *Estimated cost incurred.* This is an estimate of cost incurred, based on averages.

5. *The cost factor.* This equals the total compensation and medical cost incurred (see item 3 above) per 1,000 worker-hours of exposure:

$$\text{Cost factor} = \frac{\text{cost incurred} \times 1{,}000}{\text{total worker-hours}}$$

6. *Insurance loss ratio.* This is equal to the incurred injury cost divided by the insurance premium:

$$\text{Loss ratio} = \frac{\text{incurred costs}}{\text{insurance premium}}$$

7. *Cost of property damage and public liability costs.* This is a measure of damage to property of others caused by company operations.

8. *Nonindustrial disabling injury rate.* This is a measure of off-the-job safety:

$$\text{Rate} = \frac{\text{no. of injuries} \times 1{,}000{,}000}{312 \times \text{no. of employees}}$$

The 312 is computed as follows:

$$\begin{aligned} 7 \times 24 \text{ hours} = \; & 168 \text{ hours per week} \\ & \underline{\;40} \text{ hours of work} \\ & 128 \\ & \underline{\;56} \text{ hours of sleep} \\ & \;72 \text{ hours exposed} \end{aligned}$$

$4\frac{1}{3} \times 72$ hours per week = 312 hours per week

The above list is only a beginning. Each organization should devise an injury record system that will measure what management wants measured.

DOLLARS

Many of the measures discussed above are dollar-oriented. For internal uses, dollar-related measures are perhaps the best. Dollars are understandable and meaningful to everyone in the organization, particularly those at the corporate level. When we talk dollars, we are talking management's language. What dollar indicators can be used with management? Here are some possibilities:

1. Dollar losses (claim costs) from the insurance company
2. Total dollar losses (insurance direct costs) plus first-aid costs not paid by insurance
3. So-called "hidden costs"
4. Estimated costs
5. Insurance loss ratio
6. Insurance premium
7. Insurance experience modification
8. Insurance retrospective premium
9. Cost factor (see Chapter 4)

Losses

Dollar losses (items 1 and 2) are good indicators. (It seems more realistic to include first-aid cost than to ignore it.) Many companies use these figures or some measurements based on them, such as cost per worker-hour.

Actually, of course, the direct cost of accidents (claim costs) is not money paid out directly by the company. It is money that the insurance carrier pays to the injured employee. Hence these figures are in some sense "unreal" to management. In addition, they are not easily or accurately obtained. It may be years before actual costs of serious losses are known. Furthermore, using such figures for multiplant companies involves unfair comparison, since compensation benefits vary widely in different states.

Hidden Costs

Hidden costs are real. They consist of such items as:

- Time lost from work by an injured employee
- Lost time by fellow workers
- Loss of efficiency due to breakup of crew
- Lost time by supervisors
- Cost of breaking in a new employee
- Possible damage to tools and equipment
- Time lost while damaged equipment is out of service
- Rejected work and spoilage
- Losses through failure to fill orders on time
- Overhead cost while work is disrupted
- Loss in earning power
- Economic loss to employee's family
- At least 100 other items of cost which may arise from any accident

Although these costs are very real, they are difficult to demonstrate. To say arbitrarily to management that they amount to four times the insurable costs is asking for trouble. If management asks for proof, you can only say, "Heinrich said so." Management wants facts—not fantasy. Without proof, hidden costs become fantasy.

Estimated Costs

The actual direct cost of a work injury—compensation and medical benefits—often is not established until long after the injury occurs, especially for the more severe cases. Thus, the loss statements provided by insurance carriers do not serve the purpose of effective, *current* cost evaluation. The only information that such statements can provide on relatively recent injuries consists of the amounts of medical and compensation benefits paid to date, plus the outstanding reserves. These reserves are established primarily to assure that sufficient funds are set aside for the eventual cost of the claims. It is not until the healing period has ended and the degree of disability has been positively determined, however, that any accurate estimation of the cost can be made. And the more complicated the injury, the longer it takes to establish the final cost.

In recent years there has been an increasing demand for some method by which employers can determine promptly the approximate cost of compensable occupational injuries. Several such estimation systems are now in use in various companies in the United States. Estimated costs have proved themselves in industry to be an effective method of obtaining cost information for line measurement.

Insurance Costs

Management pays an insurance premium for workmen's compensation coverage. This is a real cost to it. How much it pays is directly dependent on the company's past and present accident record. This premium is paid directly out of management's pocket. Why not then measure the results in terms of this real out-of-pocket cost to management? The insurance premium is based on industry averages, adjusted by each company's record in the past and in some cases again adjusted by the current accident record. These adjustment factors, known as the "experience modifiers" and the "retrospective adjustment," are excellent indicators of past and present performance. Safety people ought to know, understand, and use these adjustment modifiers, since they represent the true costs of safety to management.

THE INSURANCE DOLLAR

There are both direct and indirect accident costs: direct costs are those medical and compensation costs paid to the claimant by the insurance company; indirect costs are those so-called "hidden" costs not covered by insurance and, in fact, not easily observed or recorded. Indirect costs include time lost by others who observed or gave help at the time of the accident and time lost by the supervisor in investigating.

Many studies have been made of indirect costs, with varying results. Each accident that occurs will, of course, have a different indirect cost and a different indirect-cost-to-direct-cost ratio. To attempt to strike averages for a company, for an industry, or for all industries is a meaningless exercise. Obviously, the ratio will vary tremendously, depending on circumstances, injury severity, etc.

More important, any indirect costs that a safety specialist claims on the basis of some assumed ratio, even on the basis of computed past figures, really have little validity, and management knows this. To state arbitrarily that direct costs are $10,000, indirect costs are $40,000, and total accident costs are therefore $50,000 usually is not good for much more than safety preaching.

Direct costs are more meaningful. They are close to being real costs—in fact, they *are* real costs if not clouded by insurance company "reserves." Reserves constitute the insurance company's estimate of dollar losses per case, and they must be established for all serious cases. A safety professional whose estimate shows these reserves to be inaccurate can, of course, establish an estimate with his or her own figures on those cases still open, providing, of course, that management understands this approach.

Even direct costs are somewhat unreal costs to the average company, since they are paid by the insurance carrier, not by the company itself. However, the amount the company does pay out is dependent to a large extent on what these direct costs are.

This discussion is included so that the reader might better understand how the insurance premium is arrived at—and hence might be able to use this premium in communicating with management. We are referring here only to the determination of the workmen's compensation premium, since this premium is of the greatest interest to most safety professionals. Also, in most states this premium figure is not as subject to manipulation as premiums on other lines of insurance. Thus it is a more objective measure of safety performance. The workmen's compensation insurance premium is determined by three rating systems:

1. Manual (or state) rating
2. Experience rating
3. Retrospective rating

MANUAL RATING

In making rates, the first step is to determine a basis of exposure. In workmen's compensation, the basis is payroll. This reflects the number of employees, the hours they work or are exposed, and their pay scale. It is a measure of exposure that can be more readily ascertained and more easily verified than any other.

Workmen's compensation rates are expressed in terms of dollars per $100 of payroll. Different types of work entail various degrees of hazard; therefore, in rate making, the types of work are classified to give consideration to these degrees of hazard. Through long experience and study of injury records from all types of business, a list of classifications has been developed in which the different types of work are arranged according to the degree of hazard. There are 600 to 700 classifications. Each company fits into one or more of these.

Of course, each company must first make sure that it is in the right classification, as each classification has its own rate. Each classification or group of similar industries actually sets its own insurance rate.

Each year all insurance carriers report by classification to the state and to the National Council the payroll of each company they insure and the details of losses incurred by each company for the year. Then, on the basis of what happened to all companies in the classification during the last two years, a new

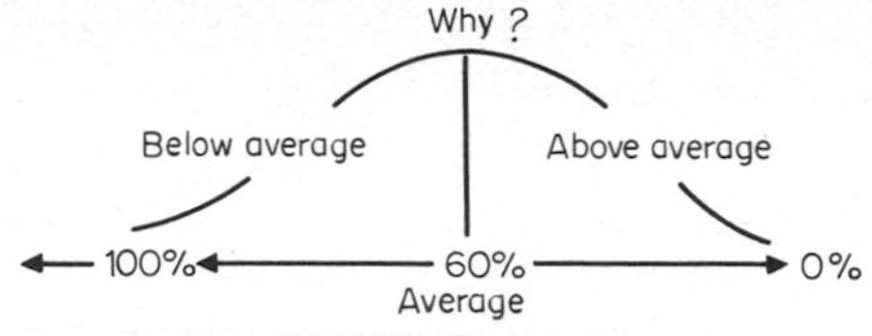

FIGURE 8-1 Normal distribution

rate is computed each year. Manual rates thus are governed by the experience of the industry over the last two years, not including the immediate past year.

After this comes experience rating. It should be pointed out that although the experience rating process described below is the one used in the majority of states, it may not be used in your state. Not all states follow this process exactly; many elect to use a somewhat different format.

EXPERIENCE RATING

After manual rates are applied, experience rating is used to vary the company's own rates, depending on its experience in recent years. Under the manual rate, all industries of the same type (such as machine shops) pay exactly the same rate, regardless of their own accident record. Obviously, in any group of machine shops, some will have a good safety record, and some will have a poor one (see Figure 8-1). Experience rating attempts to change this so that the company with a good safety record pays less than the company with high losses.

Figure 8-2 depicts the role of experience rating. Experience rating makes a statistical comparison between what losses *occurred* in a particular company during the past three years and what losses *were expected to occur* during that period in a machine shop of that size. This is done on an experience rating form (Figure 8-3).

Actual Losses

Part I of the experience rating form merely lists all accidents that have occurred in a company during the last three years. This is done as follows: First, all accidents that cost less than $750 each are lumped together. In this case:

Year 1: 15 accidents, costing a total of $4,563

Year 2: 10 accidents, costing a total of $2,288

Year 3: 10 accidents, costing a total of $1,404

Next all accidents that cost over $750 each are listed individually:

Year 1: one case still open (O) at $7,287

Year 2: one case now closed (F) at $2,587

Year 2: one case now closed (F) at $876

Year 3: one case still open (O) at $953

Year 3: one case now closed (F) at $789

These more expensive cases are then discounted. They go into the rating at a lesser amount:

Year 1: one open case at $7,287—discounted to $2,656

Year 2: one closed case at $2,587—discounted to $1,736

Year 2: one closed case at $914—discounted to $876

Year 3: one open case at $1,022—discounted to $953

Year 3: one closed case at $800—discounted to $789

This is done by a statistical formula—the larger the loss, the larger the discount. There are several reasons why the large losses are discounted in the rating plan:

1. Without this discounting, one large loss would make the cost of insurance prohibitive for the smaller company.
2. The plan states in effect that high frequency of losses should be penalized, for it indicates poor management of loss control.
3. The plan says that one serious accident is less indicative of a poor insurance risk than a batch of small ones.
4. The rating structure is still built around the theory that severity is fortuitous.

Expected Losses

In part II of the experience rating form, expected losses are computed. First, classifications are listed. (Each classification describes a type of operation; in this case, 3400 is metal goods manufacturing, 8810 is office employees, and 8742 is outside salespeople.) Next, the annual payrolls are listed, and then an expected loss rate for each classification is indicated. The expected loss rate is 60 percent of the manual rate for the classification, and this rate times the payrolls shows how many losses would be expected for that classification for that period. In the

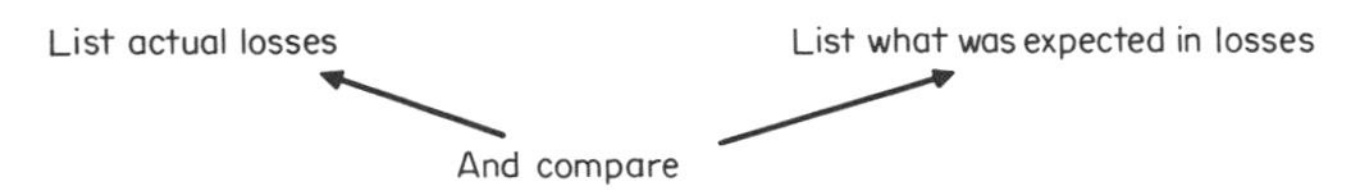

FIGURE 8-2 The experience rating process

EXPERIENCE RATING FORM

Part I – LISTING OF ACTUAL LOSSES

All cases $750 or under		All cases over $750		
		Year	**Actual loss**	**Discounted loss**
Year 1	$ 4,563	1	(O) $ 7,287	$ 2,656
Year 2	2,288	2	(F) 2,587	1,736
Year 3	1,404	2	(F) 914	876
Total	$ 8,255	3	(O) 1,022	953
Total actual losses	$ 20,865	3	(F) 800	789
Total discounted	15,265			
Difference	$ 5,600		$12,610	$ 7,010

PART II – LISTING OF EXPECTED LOSSES

Code	Year	Payrolls	Expected loss rate	Expected losses	D rate	Expected losses
3400	1	$ 696,000				
	2	724,000				
	3	741,000				
	Total	$2,161,000	1.52	$ 32,847	0.56	$ 18,394
8810	1	$ 100,000				
	2	100,000				
	3	100,000				
	Total	$ 300,000	0.05	150	0.47	71
8742	1	$ 60,000				
	2	60,000				
	3	60,000				
		$ 180,000	0.19	342	0.52	178
	Totals			$33,339		$ 18,643
	Difference				$ 14,696	

PART III – RATING PROCEDURE

Actual and expected losses	$ 15,265	$ 33,339
"B" value	7,050	7,050
Ratable excess: "W" (0.06) X $5,600	336	
1.00 – "W" = 0.94 X $14,696	13,814	
Totals	$ 36,465	$ 40,389

Experience modification =

$$\frac{\$\ 36,465}{\$\ 40,389} = 90\%\ \text{or}\ 0.90$$

FIGURE 8-3

example, we would expect in the metal goods manufacturing portion of this plant in these three years that $32,847 would go out in losses.

Then a "D" ratio, a discounting ratio, is applied. This discounts a portion of the $32,847 in losses, as we would expect a certain portion of these losses to be the result of serious accidents. The $32,847 is then discounted to $18,394. The process is repeated for the office category (8810) and the salespeople category (8742), and then totals are run.

The Comparison

We have now totaled for the company the actual losses and the expected losses. We have also discounted both. Now in part III of the form, we compare the resultant two figures. In this case, our figures are $15,265 actual losses and $18,643 expected losses. This company has a better-than-average record.

We now ask the question, "Is this believable? This is what actually happened—but could we expect it to happen again and again?" To answer this question of believability, a "B" value is added. The purpose of the "B" value (balancing factor) is to dilute our comparison—that is, to make less effective the actual comparison of our risk's experience with its expected experience. Adding this "ballast" to both sides accomplishes this. "B" values are found in a statistical table. The larger the company, the smaller the "B" value, and hence the less the dampening, the more what actually happened is used in the rates.

We then add in one final factor—the "W" factor. If the comparison is believable, as in the larger companies, through the "W" factor we will return, back into the rate, some of the excess loss that we earlier discounted. The bigger the company, the more of the severity (which we previously discounted) that will be put back into the rating formula. The "W" factor is basically only a percentage of the previously discounted losses. In our example, we are returning 6 percent of the severity. In a very large company the "B" value will become 0, and the "W" factor will become 100 percent.

Finally we merely divide:

$$\frac{36{,}465}{40{,}389} = 0.90 \qquad \text{or a 10\% credit}$$

In our example:

	Experience modification		*Manual rate*		*Payroll*		
3400	0.90	×	$2.53	×	750,000	=	$17,078
8810	0.90	×	0.08	×	100,000	=	72
8742	0.90	×	0.32	×	60,000	=	173
					Totals	=	$17,323

For the next year the workmen's compensation premium for this company will be $17,323, which is a savings of $1,924 as a result of experience rating.

RETROSPECTIVE RATING

After experience rating, the company may select a retrospective rating plan of insurance instead of a regular plan. In a retrospective plan the amount of premium paid will depend on the amount of losses that occurred this year (not in the past three years). Retrospective adjustments are made on top of, after, or in addition to experience rating.

The experience modification cannot be escaped by the company—it must pay that rate. In our example it pays 90 percent of the average, or manual, rate. With a poor loss record it could have paid 110 or 200 percent of the manual rate. When this occurs, management sometimes hopes to regain lower insurance costs by selecting retrospective rating.

Retrospective rating is a gamble—the company decides on partial self-insurance or, rather, on cost-plus insurance. The premium paid under this plan depends on the losses that the company will sustain this year. In retrospective rating, the company will pay for (1) the administration of the insurance company, (2) its losses, (3) the use of the insurance company's claim department, and (4) taxes—all subject to a minimum and a maximum. Figure 8-4 charts this.

The higher the losses go during the year, the higher the insurance premium goes. If management succeeds in keeping losses low, the insurance premium will be low. At the same time, it must pay a penalty if its losses go over the break-even point. This is shown in Figure 8-4. Obviously, under retrospective rating, management becomes quite interested in loss control.

At the beginning of the policy year management decides whether it wants retrospective rating and, if so, what kind of special plan would be best. There are a number of retrospective plans. The slope of the line, the minimum and maximum premiums, the break-even points, etc., all vary with the different plans.

As safety specialists, we should be aware of these insurance costs and of the different plans that set the costs. We should know our manual rate, our experience modifier, and whether the company is on retrospective. Management looks at these insurance costs. Safety people ought to be very familiar with these figures and should utilize them in their communications with management.

INTERNAL MEASURES

Thus, many measures can be used to judge corporate progress and performance in safety. We can use results (failure) measures such as dollar-oriented

FIGURE 8-4 Retrospective rating

measures, insurance measures, and estimated cost measures, and we can also use before-the-fact results measures such as inspections and safety sampling. OSHA code violations, discovered either by the safety specialist or by compliance officers, can also be used as measures. In short, if we want to judge a firm's internal safety performance, we have a wide variety of measures from which to choose. It is hoped that the selection will be made on the basis of some of the criteria mentioned earlier.

In point of fact, however, often top executives, rather than staff safety, choose the measures that will be used. There is nothing wrong with this, although one would hope that staff safety might encourage corporate executives to find more meaningful measures than the traditional ones.

COMPARISON MEASURES

It is less easy to find measures for use in comparing one company's progress with that of others. Most safety professionals agree that traditional methods of doing this are ineffective and inaccurate. OSHA's changeover to the Z16.1 standard probably will not alter this situation, and there are no other universally accepted measures that can be used instead of the frequency and severity rates.

Perhaps, however, the real question is whether we need such measures. Do we really need to compare ourselves with others? Does such a comparison mean anything anyway? A company's safety record reflects many things: hazards, controls, employee morale, the climate and style of the company, etc. It is difficult to see how or why a company should compare itself with others when all the items that go into the making of each company's record are different. It would seem that the only important thing we need to know is whether we are getting better or worse in each period measured.

Rather than trying to find good measures for comparing corporations, perhaps we should be searching for better ways to collect national data. This is the subject of Chapter 9. Also, we might consider comparisons based on performance measures rather than results measures. In this country we have concentrated all comparisons on results. This need not be the case. In fact, this is not the situation in other countries. In Chapter 17 we shall look at approaches used in other countries.

9

MEASURING NATIONAL SAFETY PERFORMANCE

In Chapter 5 we identified the following criteria for a measure of performance at the national level:

1. It should be valid.
2. It should be statistically reliable.
3. It should be a results measurement.
4. It should ensure input integrity.
5. It should be understandable to all.
6. It should be able to be computerized.

There is no measure currently in use which meets these criteria. Perhaps there never will be. We now use these measures:

1. Frequency rates
2. Severity rates
3. Fatalities and fatality rates
4. Estimates of costs to the country

How do these meet the established criteria? Not too well. First, they are (or could be) statistically valid and reliable, for certainly the data base is large enough. They are all results (failure) measurements. For the most part they are understandable to those who use them, and certainly they are able to be computerized. What, then, is wrong with them? The one criterion that the measures do not come even close to meeting is that of ensuring input integrity. In actuality, what we are using is batches of figures that are almost totally inaccurate, nearly to the point of being sheer guesses. There seems to be little control over the input of national figures, whether from the National Safety Council or from government. We have detailed a standard for the measures (frequency and severity rates), but that has certainly not ensured input integrity. It would be mere supposition to assume that OSHA's figures will do any better job of ensuring input integrity. Perhaps under OSHA reporting, at least the job fatality figures will be better, for all fatalities must be reported by law. But many safety professionals question whether fatalities even at the national level would be a good measure of our national safety effort.

In this chapter we shall look at some of these problems. First, however, we shall examine our current measures to see whether they could be made to ensure input integrity. Input integrity is a function less of the measurement itself than of the data collection system.

CURRENT MEASURES

As was said earlier, our current accepted national measures are the frequency rate, the severity rate, fatalities, and estimated cost measures. Measures of fatalities at the national level are valid, reliable, and understandable; they are results indicators and are able to be computerized. Their weakness lies in their lack of input integrity, of course, but even in this respect they do seem better than our other measures. Perhaps our biggest doubt comes when we ask ourselves whether a death count or death rate really reflects our national effort and performance in safety. We seem to believe that fatalities are caused largely by "luck," and thus we tend to raise our eyebrows at such a measure. And yet our criteria tell us that measures of fatalities are not too bad; perhaps they are the best of our traditional measures judging national safety performance, since they seem to be capable of more input integrity than the others.

Cost measures are a different story. While they meet all the criteria except, again, that of ensuring input integrity, here it seems that our cost figures are so far removed from reality that any attempt at using national costs is a totally ludicrous exercise. For example, the 1975 edition of *Accident Facts,* published by the National Safety Council, gives these values for the costs of accidents in 1974:

Wage loss	$13.2 billion
Medical expenses	$5.7 billion
Insurance administrative costs	$7.4 billion
Property damage (motor vehicle)	$6.5 billion
Fire loss	$3.7 billion
Indirect loss from work accidents	$6.8 billion
Broken down between:	
Motor-vehicle accidents	$19.3 billion
Work accidents	$15.3 billion
Home accidents	$5.1 billion
Public non-motor-vehicle accidents	$4.4 billion
Total	$43.3 billion

The source? The footnote to this table states the source as National Safety Council estimates (rounded) based on various inputs: national figures, state figures, and figures from insurance companies, industries, and others. How are these figures actually compiled? By piling one batch of estimates upon another batch of estimates, with almost no input of fact at all. In short, there is practically no solid, factual input whatsoever in these figures—and we all know it.

How about frequency and severity rates? These figures look so much more solid in "Accident Facts" that we think they surely must be better. They really are not. Besides the input-integrity problem, they are somewhat less meaningful to most people than dollars and deaths. But the main problem remains the simple fact that since very few people follow the standard in putting together the rates at their places of business (the input), the result, nationally, is that we add up a variety of apples, oranges, grapes, etc., and label the sum a "national indicator" of our progress in safety. It is not—it is only fruit salad.

Will this reporting improve under the required reporting systems of federal agencies? While we in the safety profession hope so, most of us remain skeptical. Perhaps the reason for our skepticism can be illustrated best by the experience of MESA (the national Mine Enforcement and Safety Administration) in attempting to generate meaningful measures and statistics from the mining organizations it works with.

In late 1975 the Department of the Interior ran an audit on the accuracy of mine operators' reports of accidents and injuries and of production and worker-hours. Each mine operator must regularly submit this information to MESA, where it is analyzed and used to determine national trends in mine safety. The 1975 audit was intended to check the accuracy of input from the mines. The Department found significant problems and inaccuracies. For example, audit tests of 105 mines indicated that there could have been 180 more disabling injuries in addition to the 283 reported by the mines—an error rate of over 63 percent. It was also found that there could have been 108 more nondisabling injuries than the 139 reported by these companies—an error rate of over 77 percent. While MESA tests indicated an error rate only one-third as great as that shown by the audit, they still indicated an error rate in nondisabling injury reporting of about 25 percent, a rather phenomenal figure.

The audit also found that of the 105 units visited, 48 incorrectly reported worker-hour data, and nine did not report any worker-hour data at all. The report stated that 26 mines had overstated by 154,741 worker-hours and that 31 mines had understated by 217,975 worker-hours; thus, it was reasoned, these errors almost canceled each other out, resulting in a less than 1 percent error in the total, or rather good accuracy. The insignificant 1 percent error actually reflects a misstatement of some 373,00 worker-hours; the inaccuracy was closer to 6 percent, which is quite significant. When we couple a 6 percent error rate in worker-hour figures with a 63 percent error rate in disabling injury figures and compute a frequency rate, we end up with an error in that figure large enough to make us have serious doubts about its validity.

Keep in mind that the above is an example of "bugs" in our data collection system. Also keep in mind that the example comes from what is probably our best federal agency in terms of enforcement of safety regulations and from a system that is probably capable of the best in data collection. I have no doubt that eventually MESA can and will collect meaningful safety data. MESA's control over the mine safety problem is infinitely better than OSHA's control over industry's safety problem. MESA people regularly contact most mines—as often as quarterly. They can iron out problems such as the one outlined above relatively easily, once they put their minds to it, for a relationship exists between them and mine management. OSHA, however, contacts only a small percentage of our nation's industrial operations, and then only infrequently. Ironing out reporting problems here is an almost impossible task. The MESA example is used only to illustrate the enormity of the input-integrity problem.

WHOM AND WHAT ARE WE ATTEMPTING TO MEASURE?

Perhaps the real key to the problem is the answer to this question: What are we trying to measure? Are we trying to measure our collective national safety performance? If so, we should be structuring our measurements along the lines followed by companies; we should be asking what percentage of companies in this country have safety policies, what percentage train their people, what percentage hold supervisors accountable, etc. Perhaps this could be accomplished with profiling systems, discussed in Chapter 17. We could sample and profile individual companies and in some way combine the results regularly into a profile of United States corporations. Our national progress, then, would be measured by the improvement in our national collective corporate profile.

Or are we trying to measure our collective national safety results? If so, we must find a solution to the input-integrity problem. Perhaps the answer is to use periodic sampling of selected plants and industries, rather than attempting mass data collection, in which input quality always varies.

While sampling covers only a small percentage of companies, statistical approaches can make this percentage quite valid and reliable and totally reflective of the national picture. More important, even with some degree of error in sampling, I have no doubt that it gives a more accurate picture than mass data collection, in which the degree of error is enormous.

Or are we trying to measure our collective national results before accidents occur? Sampling of behavior or sampling of conditions could provide these data better, with randomly selected industries sampled routinely.

What we currently are doing to find out where we stand nationally is almost totally worthless in terms of accuracy and usability. Surely we can structure a better, more practical approach.

MEASURING OSHA AND OTHER ENFORCEMENT AGENCIES

A related problem is that of measuring the effectiveness of OSHA and similar agencies charged with law enforcement and with improving safety in our nation's industries. OSHA, for example, has been under constant fire since its inception; frequency- and severity-rate indicators have shown that it has had little impact on our national safety problem. Is this how OSHA should be measured? Is this a fair measure of its performance? It seems to me to be a profoundly unfair measure.

By law, OSHA is charged with ensuring that employers provide a work environment which is free from recognized hazards that cause or are likely to cause death or physical harm and that employers comply with the standards promulgated. Thus, if we are to measure OSHA's performance, we should devise a means for measuring:

1. Whether workplaces are in fact more free from recognized hazards than they were before the enactment of OSHA
2. Whether standards are in fact compiled with

Since it is generally agreed that presence of recognized hazards and lack of compliance with standards account for a relatively small percentage of accidents (perhaps 10 percent, according to Heinrich), it is grossly unfair to charge OSHA with the other 90 percent—it is grossly unfair to measure OSHA by the accident record of this country. At best, it could impact only 10 percent of that record. OSHA was devised to get at only 10 percent of our accident problem, and therefore it should be measured by only that 10 percent of accidents with which it is concerned. Perhaps better, it should be measured by whether workplaces are freer from recognized hazards than they were before its inception and by whether standards are complied with.

OSHA MEASURES

What specifically should be measured? OSHA itself seems to like its own internal measures of number of inspections made, number of citations, and number of fines. While initially this seems to make sense, it really is nothing more than a massive numbers game. Results such as these are almost totally under the direct control of the administration. For instance, if you are being measured by number of inspections, you can improve your position merely by calling on smaller companies, which will enable you to visit more each day. Similarly, if you are being measured by number of violations found, you can easily improve your showing by calling on only a few large companies. If number of fines is what is measured, just being harder on everyone will make your record look better. These measures have almost no relation to safety and little relation to either standards compliance or fewer hazards in the workplace.

If we are going to measure OSHA accurately, we must attempt to measure exactly those things: Do workplaces have fewer hazards than before, and are standards being complied with? While the problem is not an easy one, the answer seems to lie in a sampling approach to workplaces—in sampling a statistically sound percentage of America's businesses annually and in noting progress in terms of number of hazards per employee (or worker-hour), number of violations per worker-hour, etc.

NATIONAL AND INTERNATIONAL MEASURES

Some safety professionals say that our national measures should also tell us how we stand in comparison with other nations. This would be an almost total waste of time. Different countries have different mores, values, attitudes, highway systems, cultures, job approaches, kinds and amounts of industries, job practices, job physical conditions, etc. A safety record is nothing more than a reflection of all these things. It seems illogical to attempt to compare one country's safety performance with that of another. I believe that, nationally, we should compare our present record only with our past record: Are we getting any better?

REFERENCE

National Safety Council: "Accident Facts," Chicago, 1975.

10

THE RELATIONSHIP BETWEEN MEASUREMENT AND MOTIVATION

It was stated earlier that motivation and measurement are almost synonymous at certain levels of the organization, though certainly not at all levels. In this chapter we shall examine the relationship between motivation and measurement at the various levels.

KEY MOTIVATORS AT VARIOUS LEVELS

What motivates people (or, rather, what causes people to motivate themselves) at various levels? Figure 10-1 attempts to describe the process of motivation. At various levels of the organization, people are "turned on" and "turned off" by different things. These "key motivators," listed in the left-hand column, are based on certain theories from the behavioral sciences (as indicated in the right-hand column).

Employee Level

At the employee level, the primary motivating influence is peer pressure, or pressure from the informal group; this is described in some detail in Chapter 13. Other key influences on employee attitudes and behavior (actions) are whether the job is important and meaningful to the employee, the degree of employee involvement and participation in the important aspects of the job, and how much recognition is given for good performance. These insights come from the thinking of Frederick Herzberg (see Chapter 13), Chris Argyris, Rensis Likert, and other influential management theorists.

Conflict Theory Argyris's theory provides safety managers with insights into why people commit errors. Argyris takes human nature as a starting point and analyzes the process of growing up and maturing. Children, he says, are passive and dependent upon their parents; they exhibit few behaviors. Their interests are shallow and short-term. They are at all times in a subordinate position in

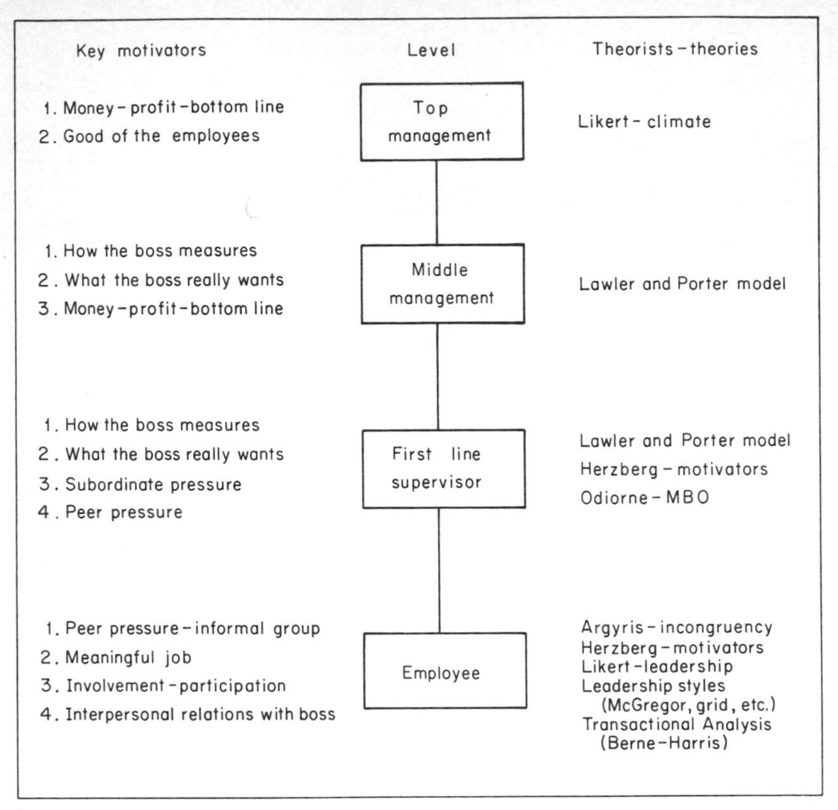

FIGURE 10-1 Motivators at various levels

their relationships with older persons, and they are relatively lacking in self-awareness. As people mature, this changes. Adults are active, independent creatures who like to stand on their own. They exhibit many behaviors, and their interests are deep and long-term in nature. Mature people view themselves as equals in most relationships, not as subordinates, and they have self-awareness.

According to Argyris, all organizations—whether industrial, governmental, mercantile, religious, or educational—are structured under certain principles:

1. They have a *chain of command.* This creates a superior-subordinate relationship, which in turn causes workers to be dependent on the boss, to become passive, and to lose interest in the job worker.

2. The *span of control* is small. This creates dependency and reduces the freedom and independence of the worker.

3. There is a *unity of command,* or only one boss. This creates dependence and highlights the subordinate role of the worker.

4. They are characterized by *specialization.* The work is broken down into small simple tasks, which leads to a lack of interest, to a lack of self-fulfillment and feelings of self-importance, and to dependency and passivity.

These principles create "managerial pyramiding," in which there is a boss, a series of minor bosses under that boss, and a series of minor-minor bosses under them, down to the workers, who would like to consider themselves equals but find that they are very *unequal.* Such structures make workers highly dependent on their bosses.

Argyris says that these principles of management are, in fact, in conflict with the needs of individuals, and he suggests that this conflict causes people to quit, to become apathetic about their jobs, to lack motivation, to lose interest in the company and its goals, to form informal groups, to cling to the group norms instead of to the company's established norms, and to evolve a psychological "set," believing that the company is wrong in most things it attempts to do. It also causes accidents as a result of inattention, disregard of safety rules, poor attitudes toward the company and safety, etc. The normal management reaction to these symptoms is more control, more specialization, and more pressure.

What can we do about this? Since we cannot feasibly change mature people into immature ones (nor would we want to), the only option is to look at the organization and see how we can change it. This leads us to organizations (and safety programs) in which there is less control, less pressure, fewer superior-subordinate relations, etc.

Argyris, one of the most famous of the behavioral scientists, proposed "leveling." The use of group decision making and supervision means that the boss does not necessarily make all decisions alone. The emphasis here is on the involvement of people in the decision-making process to the extent that their perceptions of problems are sought, their ideas on alternative solutions are cultivated, and their thoughts on implementing decisions which have already been made are solicited.

Motivation-Hygiene Theory The second behavioral theory we shall consider is Herzberg's motivation-hygiene theory. Herzberg calls certain variables "hygiene factors" and others "motivation factors." By improving the hygiene factors (company policy, supervision, interpersonal relations, status, etc.), we can make a dissatisfied worker into a satisfied worker, although this does not necessarily mean that the worker will be motivated. The motivation factors (achievement, recognition, the work itself, responsibility, etc.) have to do with the job itself, while the hygiene factors are peripheral to the job.

The following things determine the worker's level of satisfaction:

1. Money

2. Status

3. Relationships with the boss
4. Company policies
5. Work rules
6. Working conditions

The following factors determine the worker's level of motivation:

1. Sense of achievement
2. Recognition
3. Enjoyment of the job
4. Possibility of promotion
5. Responsibility
6. Chance for growth

If we structure the safety program on the basis of the things that relate only to satisfaction, people will never be excited by it. They probably will be bored by it. On the other hand, if we build the safety program around the things that motivate, it has a good chance for success.

Likert's Theory The third theory we shall discuss is that of Rensis Likert, whose studies concerned the effect of the supervisor-employee relationship on productivity. Among Likert's findings are:

The tighter the supervisor's control over the employee, the lower the productivity.

The more the supervisor watches and supervises the worker, the lower the productivity.

The more punitive the supervisor is when the employee makes a mistake, the lower the productivity.

In short, Likert's research indicates that if you want productivity, you should not "control" employees. Controlling employees will cause them to work less.

Today management accepts these behavioral theories, and certainly they can help safety directors increase the acceptability and productivity of a safety program.

Thus as we examine the relationship of motivation to measurement at the employee level, we find that it is not strong; in fact, there could even be some negative correlation. The more we tend to measure and control employees, the less we are likely to get from them. At the employee level motivation is better achieved by means of peer pressure, by treating employees as mature and intelligent human beings, and by not controlling them.

We have not yet been very successful in applying these theories to safety management. We have not really "turned employees on" to our safety programs. You may wish to find out how successful your program has been by asking employees what it means to them. Often a safety director who does this finds that the employees think the safety program consists of the list of work or safety rules on the bulletin board, of some silly posters, of boring safety talks that must be periodically endured, etc. These are typical reactions to a traditional safety program.

Younger versus Older Workers Perhaps safety programs like these were adequate in the past, but today they are not, especially with many younger workers. There is a definite difference between the personal values and attitudes of the younger worker and those of the older worker. This is important to safety professionals, who must sell their safety programs to both older and younger workers.

Why are younger and older workers different? They are different because their values and attitudes were determined by different institutions and different lifetime experiences. Consider the changes that have occurred in recent years in the three basic institutions of the family, the church, and the school:

1. The family now has considerably less influence. More women work, more men travel, and more families have become fragmented, as family members move far from one another.
2. The church has less influence and appears to be losing ground steadily.
3. Schools have changed their teaching methods, and they too appear to be losing influence. Old teaching methods based on memorization have given way to the concept of encouraging students to think through a problem for themselves.

Furthermore, the war in Vietnam, the nuclear threat, the civil rights movement, concern over the ecology, and the communications explosion have all contributed to the change in values and in attitudes. The young today have a distaste for social and institutional rigidity. They fear the depersonalizing effects of technology. They are intolerant of hypocrisy. They have different life-styles and a different work ethic. The credo of the older generation was that we live in order to work. The younger generation believes that we work in order to live. Younger workers do not like doing things the hard way if there is a simpler way. Often they resent it when the boss says, "Do it my way." They often balk at safety rules. They believe in being able to "do their own thing."

Although younger people rely on institutions to do the things they cannot do themselves, at the same time they have lost confidence in the large institutions of government and business. Where does this leave us? It leaves us with people who have different values and attitudes from the ones we are used to. We are not going to change these people or their value system. We have to accept them as they are and learn how to live with them. We shall also have to reconsider our styles of leadership and our training approaches. Future managers will be using more employee-centered leadership approaches, more participative styles of management, etc. Safety training in the future will center more on understanding and dealing with people than on teaching technical subjects. It is common in safety training today for supervisors to teach transactional analysis and other means of understanding people better.

First-Line Supervisory Level

At this level of the organization there are major differences in what seems to motivate employees. First-line supervisors have moved up a step in the organization, and they are now in a totally different situation psychologically. Whereas they used to be motivated primarily by peer pressure, this is now much less important. Members of the old peer group are now supervisors themselves. The group is less strong, and its pressure is less important. Now there is more interest in the boss and in his or her desires, for pleasing the boss means retaining the new position of power over others and perhaps even being promoted again to a position of more power and larger financial gains. Thus the first-line supervisor's primary motivational force is the boss—how he or she is being measured by the boss, how to demonstrate good job performance to the boss, and how to find out what the boss really wants. The measures the boss uses are not always a good indicator of what the boss really wants, however, as many of them have been chosen by management. Also, priorities shift and change, without there being a concurrent change in the measures the boss uses. Thus first-line supervisors strive to find out what is important to the boss, and they react to those itcms first.

It is true that first-line supervisors also react to pressures from other directions. Since they must live with their people—their subordinates—they do react to them and to the pressure they exert. And there is still a small amount of peer pressure, but this usually is not a large factor in the first-line supervisor's behavior. The first-line supervisor is motivated first by what the boss wants (as indicated by the measure used and by the boss's expressed wishes), and second by the members of the group. The first-line supervisor must know these people well enough to know what they want and need; success on the job depends on his.

These motivational pulls are described well by a number of behavioral scientists and management theorists. The model of supervisory performance devised and tested in full by Edward Lawler and Lyman Porter is one of the most descriptive; it is discussed in Chapter 12. Also discussed in Chapter 12 is the way the motivation-hygiene theory of Frederick Herzberg might pertain also to supervisory motivation. Figure 10-1 lists one other theorist, George

Odiorne, whose ideas seem to apply at this level. Odiorne has written extensively on the concept of management by objectives (MBO). This is perhaps one of the best approaches to supervisory motivation. It has also been used effectively to elicit supervisory performance in safety which had previously been missing. This was discussed briefly in an earlier chapter.

Middle-Management Level

What motivates middle managers is very similar to what motivates first-line supervisors, with one major exception. While they are very much interested in the measures of their performance used by the boss and in the desires and wishes of the boss, middle managers are especially responsive to measures relating to the dollar—to profitability, which is so important to managerial thinking. The Lawler and Porter model discussed in Chapter 12 is also relevant at this level, the only difference being that results measures are probably better used as indicators of performance and effort at this level than at the lower level. While not indicated on the chart in Figure 10-1, the MBO approach is quite effective and relevant here.

Top-Management Level

What motivates top management? Top-management people are highly interested in money, in profitability, in the bottom line; they must be, for that is their primary responsibility in the organization. However, those in top management often are also deeply interested in the good of the people who work for them. Although not many theories relate directly to the top levels of the organization, Rensis Likert's work on corporate climate is relevant here. After making an in-depth study of the impact of leadership styles on productivity, Likert began studying the impact of total corporate style on productivity. His research led him to study the following factors in organizations and to help organizations improve their corporate climate by making improvements in these areas:

1. Whether there is a feeling of mutual confidence and trust on the part of employee and management

2. The amount of interest shown by management in the future of subordinates

3. How much understanding management has of subordinates' problems

4. The amount of training and help that is given

5. Whether subordinates are taught to solve problems or are merely given answers

6. The amount of support given with physical resources

7. How much information is given to subordinates that they want, as opposed to information that they must have
8. How much employees' ideas are sought out and used
9. How approachable management is
10. How much credit and recognition is given

Executives are generally deeply interested in building a good climate within their companies. While profitability must remain a top priority to them, most executives are almost as interested in building a good corporate climate, for they know future profitability.

THE MEASUREMENT-MOTIVATION RELATIONSHIP

Thus we find a very tight relationship between motivation and measurement at the first-line supervisory level and the middle-management level, while other factors are of more motivational importance at the other two levels. For the first-line supervisor and for the middle manager, however, the number one motivator is the measurement system that is used.

REFERENCES

Argyris, Chris: *Personality and Organization,* Harper & Brothers, New York, 1957.

Berne, E.: *Games People Play,* Grove Press, Inc., New York, 1964.

Blake, R. and **J. Mouton:** *The Managerial Grid,* Gulf Publishing Company, Houston, 1964.

Harris, T.: *I'm OK—You're OK,* Harper & Row, Publishers, Incorporated, New York, 1971.

Herzberg, F.: *Work and the Nature of Man,* The World Publishing Company, Cleveland, 1966.

Kollatt, D. and **R. Blackwell:** *Direction 1980: Changing Life Styles,* Columbus Management Horizons, Columbus, 1972.

Likert, R.: *The Human Organization,* McGraw-Hill Book Company, New York, 1967.

McGregor, D.: *The Human Side of Enterprise,* McGraw-Hill Book Company, New York, 1970.

Odiorne, G.: *Management by Objectives,* Pitman Publishing Corporation, New York, 1965.

Porter, L., and **E. Lawler:** *Managerial Attitudes and Performance,* The Dorsey Press and Richard D. Irwin, Inc., Homewood, Ill., 1968.

PART FOUR

MOTIVATING SAFETY PERFORMANCE

During the past 65 years, industrial safety people have tried to motivate employees with a deluge of slogans, posters, contests, displays, meetings, films, payroll inserts, letters to their homes, charts, literature, and assorted gimmicks. In other words, we have relied primarily on publicity to spur employees toward safety.

Publicity is certainly nothing to be sneered at. The pen is powerful. The written word has altered the course of history and reshaped the world many times. But always, when this happened, the words were aimed at a receptive audience. They were written for people who were willing to listen to them and act on them.

Voltaire and his fellow writers, for example, smashed the tyranny of church and state in eighteenth-century France and partly formed the world that we still live in today. But what would Voltaire's deadly wit have accomplished if he had lived and written in the Middle Ages? Probably nothing but a stake and a fire in a public square—with his own body serving as the fuel.

Too many of us in the safety profession have been playing the part of a Voltaire trying to reform the Middle Ages. All our brilliant propaganda in behalf of safety has fallen on uninterested ears and bored minds. The people in the plant are not receptive. Too often, we motivate nothing but a "ho-hum."

Yet we who want to change the behavior of the employees of one company have a great advantage over those who are trying to influence the general public. Company management has the power to directly alter the attitude of its employees to make them receptive to our posters, literature, films, etc. Management does this every day to get what it wants from its employees—except in accident prevention.

The next few chapters focus on the various ways and means available to us for provoking action in the three categories of people we work with: employees, supervisors, and executives.

11

MOTIVATING MANAGEMENT

We have no guidelines—that is, no motivational research—to help us determine the best ways to motivate management (incite it to action). Ideally, we should not have to motivate management; management should actually be doing the motivating. Unfortunately, this is not usually the case today, and most safety professionals find themselves in the position of having to motivate their top executives. Often they find that this is their most difficult motivational task.

Perhaps we should first examine our own beliefs and feelings in regard to top management. The above paragraph, I believe, presents an accurate assessment of the way most safety professionals feel about the situation—that our most difficult motivational task concerns the boss. When we say this we are saying that we believe the boss is opposed to what we are trying to achieve in the organization—that the boss is opposed to safety, or at least opposed to our approach to safety. We are saying that we must motivate the boss to be in favor of what we are trying to do.

Safety professionals should, I believe, examine this kind of thinking in the light of reality. Are top executives against safety? Do they really want their people to get hurt? Obviously, they do not. Such an attitude would be the direct opposite of the way most executives feel about creating a good corporate climate, discussed in the last chapter.

WHAT THE EXECUTIVE IS LIKE

While there has been little research on motivating executives, studies of executive characteristics have been made. These studies are of value to us, for they do not describe a person who is uninterested in the safety of others. Executives have been shown to have the following traits:

1. They are usually family-oriented and have happy homelives.
2. They are educated and intelligent.

3. They are interested in further self-development.
4. They have many outside interests.
5. They are usually religious.
6. They enjoy good health.
7. They are vitally interested in people.
8. They take risks only after thorough study of the facts.
9. They are greatly interested in new and different approaches to problems.
10. They are objective in their approach to problems and straightforward in their relationships.

This is not the profile of a group of people who do not care whether their employees get hurt. This profile describes individuals who are interested in the safety of subordinates and who are willing to invest personal time and corporate money, to achieve safety. Why, then, do professional safety people have such a difficult time selling their ideas to the people who really want to solve a company's safety problems?

We should begin by believing that executives do want safety and that we may have had difficulties with them in the past simply because they did not buy a particular solution to a problem. In the past, executives may have been less than enthusiastic about our traditional safety solutions, and perhaps they were right to feel this way. If we had been selling anything else with those kinds of approaches, chances are that top executives would not have bought that either. For instance, no top executive would buy an approach to production that was based on committees, posters, and preaching.

We go a long way toward solving our problem if we take the point of view that we do not have to sell the boss on the product (safety), only on the solution. Then all we have to do is find a solution which will be acceptable to the boss and which will work. We can forget about our traditional approaches to safety, since they have proved to be neither acceptable nor successful in most cases, and concentrate on the many other approaches and strategies we can use that do meet these two simple criteria. There are systems approaches, MBO approaches, training approaches, team approaches, etc. Executives have not rejected these approaches to getting things done in other important areas, and chances are they will not reject them when it comes to safety.

Perhaps the most important exeuctive characteristic of the ten listed above is the tendency to be interested in new and different approaches. If the safety professional's solution embodies a new and different approach, it is enhanced in the boss's eyes. As noted in the list, executives generally take risks only after a thorough study of the facts. Thus the safety professional should have all the facts ready for the boss.

THE ANNUAL REPORT

Every safety professional ought to utilize the tool of the annual report to motivate management. Often management does not ask for such a report, but it should be submitted anyway. It would certainly be peculiar for management to appoint a person to handle the staff safety function (whether full or part time) and then never ask how that person is doing—or what the company is getting for the money spent. Thus, although management may not ask for a report, the safety professional should give it!

Whatever the safety professional wants management to know should go into the annual report. For example, the report should answer the following questions:

1. How did we fare last year? (Give results expressed in management's terms.)
2. What did we accomplish last year? In what ways are we stronger than we were before?
3. What are our objectives for next year? How will we be stronger at this time next year?
4. What do we need from management? How is it letting us down?

This report is crucial to a safety professional's relationship with management. The goals you set for the future, as indicated in the report, tell management in what direction you intend to move. And if management approves, it is committed to that direction.

Only two motivational techniques are discussed in this chapter—using the dollar and submitting an annual report. Other effective techniques are part of the safety professional's everyday working relationship with the executives. Demonstrated competence will be responsible for much action by management.

MANAGEMENT MOTIVATORS

As noted in the last chapter, profitability of the organization and the well-being of employees are of primary concern to top management. These, then, are the keys to the safety professional's relationship with executives. Management is interested mainly in the relationship of the safety professional's ideas to the profits of the organization. That is, what will management get in return for the money it is being asked to spend? Thus we ought to be dollar-oriented when we talk to management. Even if management understands the language of frequency and severity rates, we ought to talk in terms of dollars.

The second major motivator at the executive level is what is good for the employees of the organization. This is, of course, a natural for safety. In fact, the connection is so obvious to executives that we hardly need to preach on the subject.

GETTING TO THE EXECUTIVE

Perhaps our major problem in dealing with executives is accessibility—getting to them when we need to. One study of executive behavior discussed the so-called "diary complex." Executives are ruled by their appointment calendars. They do what the calendar says; they see people who are listed theron. The appointment book is often controlled by someone else, such as an assistant or a secretary. Thus the problem of gaining access is often a problem of getting on the list, which in turn may depend on knowing who controls the schedule. This, however, is a problem only the first time you try to see the boss. Whether you will get to see the boss on a regular basis will depend on what you do during this first meeting. The following points are important:

1. Talk in management language (in terms of dollars and employee well-being).
2. Say something that is of importance to the boss. For example, discuss progress being made and specific problems you have uncovered.
3. Make it clear what you need and want from the boss. For instance, say "I want to send you a monthly report that will indicate which supervisors have performed well in safety and which have not. Then I would like you to call three of these people on the phone each month either to compliment them or to complain about their performance."

Obviously, each executive is different and must be handled differently. Your job is to get to know the executive with whom you will be working.

REFERENCES

Carlson, S.: *Executive Behavior,* Strombergs, Stockholm, 1951.

Wald, R. and **R. Doty:** "The Top Executive: A First Hand Profile, Skills That Build Executive Success." *Harvard Business Review,* September 1965.

12 MOTIVATING SUPERVISORS

The attitude of the majority of supervisors today lies somewhere between total acceptance and flat rejection of comprehensive accident prevention programs. Most typical is the organization in which line managers do not shirk this responsibility but do not fully accept it either and treat it as they would any of their defined production responsibilities. In most cases their "safety hat" is worn far less often than their "production hat," their "quality hat," their "cost control hat," their "methods improvement hat," etc. In most organizations, safety is not considered as important to the line manager as many, if not most, of the other duties that he or she performs.

On what does a manager's attitude toward safety depend? It depends on (1) abilities, (2) his or her role perception, and (3) effort (see Figure 12-1). All are important, and a manager will not turn in the kind of performance we want unless we take all three into account.

EFFORT

Two basic factors determine how much effort a person puts into a job: (1) his or her opinion of the value of the rewards and (2) the connection the person sees between effort and those rewards. This is true of a manager's total job, as well as of any one segment of it, such as safety.

The Value of Rewards

The manager looks at the work situation and asks, "What will be my reward if I expend effort and achieve a particular goal?" If the supervisor considers that the value of the reward which management will give for achieving the goal is great enough, he or she will decide to expend the effort.

"Reward" here means much more than just financial reward. It includes all the things that motivate people: recognition, chance for advancement, increased pay, etc. Most research today into supervisory motivation indicates that the rewards of advancement and responsibility are the two greatest motivators.

In effect, the line manager looks at the rewards that management offers and makes a judgment as to whether those rewards are great enough. If the rewards consist of advancement and additional responsibility, rather than some of the

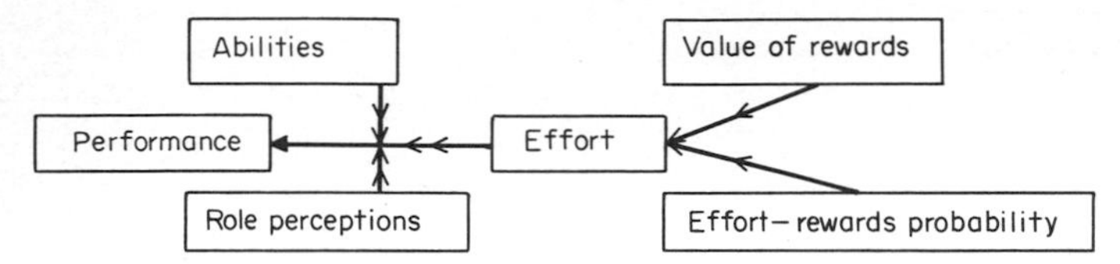

FIGURE 12-1 Factors affecting performance

lesser enticements that management too often selects, the manager will probably decide that they are worth the effort. In safety, as in other areas, the rewards that management chooses to give are often too small or too unimportant to entice the line manager.

The Effort-Rewards Probability

In assessing whether rewards really depend upon effort, the manager asks the following kinds of questions:

1. Will my efforts here actually bring about the results wanted, or are factors involved that are beyond my control? (The latter seems a distinct possibility in safety.)
2. Will I actually get that reward if I achieve the goal?
3. Will management reward me better for achieving other goals?
4. Will it reward the other manager (in promotion) because of seniority, regardless of my performance?
5. Is safety really that important to management, or are other areas more crucial to it right now?
6. Can management really effectively measure my performance in safety, or can I let it slide a little without management's knowing?
7. Can I show results better in safety or in some other area?

The line manager asks these questions and others unconsciously before determining how much effort to expend on safety. He or she must get the right answers before deciding to make the effort needed for results. In safety such questions are crucial. Often line managers decide that their personal goals would be better achieved by expending efforts in other areas, and too often their analysis is correct because management *is* rewarding other areas more than safety.

Changing this situation is the single greatest task of the safety professional. Change can be achieved by instituting better measurement of line safety performance and by offering better rewards for line achievement in safety.

ABILITY

Job performance does not depend simply on the effort that managers expend. It depends also on the abilities they bring to the task. The word "abilities" refers here both to inherent capabilities and to specialized knowledge in the particular field of endeavor. In accident prevention, this means that we must ensure, through supervisory training, that line managers have sufficient safety knowledge to control their people and the conditions under which these people work. In most industries lack of knowledge is not a problem. Usually, however, line managers know far more about safety than they apply. Many people agree that managers can achieve remarkable results on their accident records merely by applying their management knowledge, even if they have little safety knowledge. If a manager does not have adequate safety knowledge, the problem is easily handled through training.

ROLE PERCEPTION

Role perception is even more important than ability. Line managers' perceptions of their role in safety determine the direction in which they will apply their efforts. Lawler and Porter describe a good role perception as one in which the manager's views concerning placement of effort correspond closely with the views of those who will be evaluating his or her performance.

In safety, role perception has to do with whether line managers know what management wants in accident control and with whether they know what their duties are. In the area of role perception, the safety professional should search for answers to some questions about the organization and about each line manager in it. These questions concern the content and effectiveness of management's policy on safety, the adequacy of supervisory training, company safety procedures, the systems used to fix accountability, etc.

A MEASUREMENT SYSTEM

It is conceivable that a measurement system could be devised to measure a line manager's safety performance in terms of ability, role perception, and effort. Perhaps such a system would enable management to decide what level of safety achievement should be expected for a given manager. Such a system of measurement would have to be based on self-rating and/or rating by management. A system of rating the three elements that go into safety performance (in terms of points, percentage points, or whatever seems logical) would help to identify supervisory performance in safety before the fact—before the accident occurs. Perhaps some equation could be devised:

$$\text{Effort} \times \text{ability} \times \text{role perception} = \text{safety performance}$$

It is possible that rating systems could be devised for each of the three factors, resulting in a final number which would be an indicator of a manager's safety performance.

MOTIVATORS AND DISSATISFIERS

In Chapter 13 we shall look at some research findings from the behavioral sciences concerning what motivates and dissatisfies employees on the job. The findings are also relevant at the supervisory level.

The four most important motivators have been shown to be:

1. Advancement. Advancement seems to be the most important motivator for supervisors, considerably more so than for employees.
2. Responsibility. Degree of responsibility is important, but less important than in the case of employees.
3. Possibility of growth. Possibility of growth is far more important to supervisors than to employees.
4. Achievement. Achievement is important to supervisors, but not as important as it is to employees.

What dissatisfies supervisors is similar to what dissatisfies employees, except that "quality of work," the major employee dissatisfier, has been shown not to be a factor at the supervisory level.

Application of Theory

If we are to apply this knowledge of motivation, we might consider:

	Motivator
1. Making safety performance an integral and important part of supervisors' performance	Advancement and responsibility
2. Giving supervisors a free hand in how they control accidents, retaining accountability for results	Responsibility growth
3. Assigning supervisors special projects in safety	Growth and responsibility

Training

As with employees, training is a source of motivation for supervisors, in addition to providing needed knowledge and skills. There are many sources of help for supervisory training, including packaged courses, self-study courses, programmed courses, and discussion courses. They are offered by universities, vocational schools, government agencies, insurance carriers, trade associations, and other institutions.

Some courses are excellent, some are good, and some are poor. The selection must be made carefully, and the content and method of the courses under consideration must be screened carefully. Some may be a waste of time. The primary question in selecting outside help should be whether the intended program will fit your defined objectives. The first step is to determine whether training is the answer to your defined problem; if it is, the second step is to define the objectives of that training.

If a broad base of knowledge in accident fundamentals is needed by your supervisors (perhaps only the newly appointed ones), the self-instruction programs on the market might help. Generally, however, training is most successful when it is conceived and presented (or at least introduced) by management. This gives the training the aura of importance that it needs in order to be effective.

REFERENCES

Lawler, E. E., and **L. N. Porter:** "Antecedent Attitudes of Effective Managerial Performance," University of California, Institute of Industrial Relations, Berkeley, 1967.

Myers, M. S.: "Who Are Your Motivated Workers?" *Harvard Business Review,* January–February 1964

13
MOTIVATING EMPLOYEES

In this chapter we shall concentrate on the employee. How do we motivate the worker to be safe? No one has the magic key to understanding all people—and hence to knowing how to manipulate them to want to do what needs to be done. All we can do is attempt to gain some insights. We shall look at some influences that help to create employee attitudes—both those influences which we can consciously apply and those over which we have little or no control. These influences help to mold and shape the employee's decision concerning how he or she will work. The worker makes this decision; management cannot. Often, however, management's definition of policy influences the employee in making the decision.

Let us imagine a factory worker named Elmer who slaves over a machine all day to produce 275 Super Speed Fishing Worm Untanglers, size 4. One day management hands out booklets telling Elmer and his fellow workers how they can produce 300 untanglers instead of 275. Nobody *enjoys* making fishing worm untanglers, and Elmer skims through the booklet, throws it away, and keeps right on turning out 275 a day. Management then sends out a warning to the effect that any employee who fails to produce 300 untanglers a day will be dismissed. Elmer is now powerfully motivated. He finds that the booklet is interesting reading after all, and he learns everything in it with remarkable ease.

This is more typical of the way management motivates its employees to do what it wants in production, cost control, and quality control than in accident prevention. In these other areas, management does not seem to worry as much about motivation. It decides what it wants done, and then it makes sure that this *is* done. In safety, we seem to be more concerned with motivation. We decide what we want, and then we use contests, posters, meetings, and special campaigns to persuade employees that they ought to do it.

How does management get other things that it wants—production, for instance? When management officials decide they want a certain level of production, they:

- Say what they want. They *communicate.*
- Say to someone, "You do it." They *assign responsibility.*
- Say, "You have my permission to do whatever is necessary to get the job done." They *grant authority.*
- Say, "I'll measure you to see whether you are doing it." They *fix accountability.*

We can use some of these principles in safety also. This does not mean, however, that we can force employees to work safely. But we can set a stage where it is easier for them to decide to work safely.

Management, through its policy, makes the decision that safe performance from employees is desirable. Management cannot, however, force a safe performance. Employees decide for themselves whether they will work, how hard they will work, and how safely they will work. Their attitudes shape their decisions—attitudes toward themselves, their environment, the boss, the company, and their entire situation. Employees' decisions are based on their knowledge, their skills, and their group's attitude toward the problem.

All we can do is recognize those influences over which we have no control and extend our influence wherever possible. Some of these influences are:

1. The influence of the attitudes of the group toward safety
2. The influence of selection and placement
3. The influence of training
4. The influence of supervision
5. The influence of special-emphasis programs
6. The influence of media

We shall discuss each of these individually.

THE GROUP

Although each employee is an individual, he or she is also an integral part or a member of a group, and every manager must take this into account.

In chemistry, elements combine to form substances that have entirely different properties from those of the individual elements. People combine to produce groups that have entirely different properties from those of the individuals. We have to recognize these group properties, just as we do the individual's properties. Each group has a distinct personality of its own.

Each group makes its own decisions. It sets its own work goals, which may be identical with management's goals or different. Each group sets its own moral standards. For instance, a group might decide that stealing things from the company is allowable and hence would not exert group pressure on the individual who takes small items (pencils, notebooks, hammers, etc.). However, a group member who is caught stealing from another group member is usually in deep trouble. The group takes care of the punishment itself. Management does not have to because the group's moral standards state that it is wrong to steal from another group member. The group has decided, and the group enforces its decision.

The group also sets its own safety standards, which it lives by, regardless of what management's standards are. Take, for example, hard hats. If a group of construction workers make the decision that hard hats should be worn, all members will wear hard hats. If, however, they decide against head protection (which they did some years ago), they will exert pressure on members not to wear it.

How can we cope with the phenomenon of the group? First, we try to understand the groups that we have in our company; second, we try to identify the respective strengths of those groups; and third, we try to build strong groups with goals that are the same as our goals in safety.

What Makes a Group?

There is only one real criterion for a group: The members must have the same goals. If a group's goals are important to it, the group will be strong; conversely, weak goals mean a weak group. An individual member of the group who does not share the group goals becomes a weak member of the group. For each group the manager must determine two things: (1) Is it strong or weak? (2) Are the group goals compatible with company goals?

We can determine the strength of groups by bearing in mind some characteristic symptoms of strong and weak groups. In a strong group, the members *voluntarily:*

1. Try to deserve praise from the rest of the group
2. Seek recognition from the group leaders
3. Exert pressure on weak group members
4. Put special efforts into achievement of the group goals

In a weak group, the members:

1. Form cliques or subgroups
2. Exhibit little cooperation
3. Are unfriendly
4. Use no initiative

5. Avoid responsibility
6. Have no respect for group policies

Building Stronger Groups

Before we can build an effective group with goals identical to ours, we must strive to build up the group's strength. Four things are essential for a strong group:

1. Individual competence. Each member of the group must have the ability to pull his or her own weight.
2. Individual maturity. Each member must be mature. A strong group dislikes members who will not do their share. It dislikes those who exert more effort pleasing the boss than the other group members. A strong group dislikes the "let-George-do-it" type.
3. Individual strength. Each member must have not only ability and maturity but also the strength to do his or her job and to earn group respect. This means no "weak sisters" and no "loners."
4. Common objectives. Common objectives, as we have said, are essential to a strong group.

The influence of the group is one that we cannot control. It is, however, an influence which we must understand and which we can bend slightly to our purposes. For example, we can try to ensure that our individual workers are competent, mature, and strong by utilizing effective selection and placement procedures.

THE INFLUENCE OF SELECTION AND PLACEMENT

This very basic control has in recent years become increasingly difficult to utilize. Government today has a large amount of control over the process, and some industries have almost abdicated control in this area. However, even if we cannot fully ensure good selection processes, we certainly ought not to abdicate control in this area entirely. It is far too important to give up. For example, even in the most difficult situations, management can check with previous employers to find out something about an applicant's history.

Figure 13-1 gives a broad outline of the process management can use in selecting employees. Ideally, the process is based on job standards which state that for a particular job a particular type of person is required. This is utopian, and the "exactly right" person for the job is seldom (if ever) found, but such a goal is still worth striving for. As the figure shows, we have two basic sources of information about an applicant: (1) biographical data and (2) test results. Biographical data are by far the more important. Past performance is still the

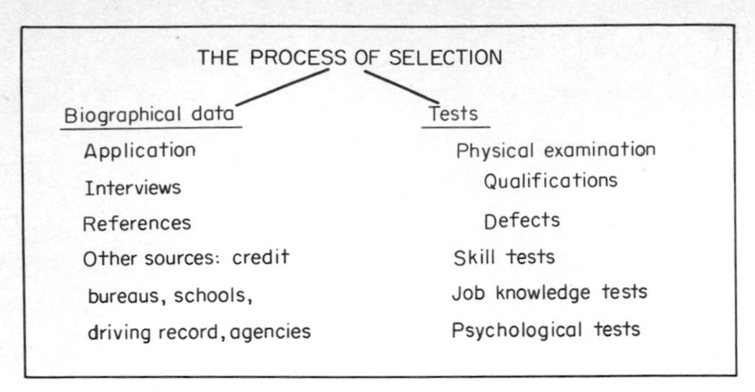

FIGURE 13-1 The selection process

best indicator of future performance. A person probably will work tomorrow in a manner similar to the way he or she worked yesterday. Tests are secondary (with the possible exception of the physical examination in certain industries). They are important and should be studied for possible use, but only after the company has done an excellent job in the area of biographical information.

If we utilize the tool of selection properly, not only can we ensure that our workers will have relatively few accidents, but we can also try to get workers who will build strong groups for us—workers who will become competent and will exhibit maturity and strength.

THE INFLUENCE OF TRAINING

Training is a powerful influence and motivator in safety, just as it is in many other areas. In training employees, you give them only two things—knowledge and skill. You get three things back—knowledge, skill, and motivation. This has been shown over and over again in business; the value of training has been proved. Industry today is engaged in more training that ever before in history.

Training is, however, often terribly misused; it is too often considered a panacea for problems in industry. Too often management perceives a performance problem and, without diagnosing the problem, decides that some training is needed, when the problem may actually be in poor procedures, poor selection, or poor management, for example.

Training is effective when it is aimed at defined needs—when analysis shows the problem to be lack of knowledge, lack of skill, or even lack of motivation. It is a waste of time when the problem is elsewhere in the system. Much has been written on how to train, whom and when to train, and what to provide training in. Obviously, the subject of training is too big and too involved to cover in a few short paragraphs. There are a few principles we ought to discuss, however.

How to Train

Training used to be a simple four-step process: tell, show, observe, and correct. Industrial training today is much more complex than this, but it does start with one essential step: Define the objectives of your training. Objectives should *always* be spelled out before content and methods are decided upon. Training objectives should be stated in terms of desired behavior. They should state how the participants should perform when they have completed their training. For instance, as a result of this course, the participants should be able to:

1. Investigate any injury accident to determine five causes and record the information properly on the accident investigation form
2. Transmit management's policies on safety to employees
3. Orient each new employee in safety
4. Make a job safety analysis of each job

Action, not knowledge, is the purpose of training. This philosophy is most important. Training in safety, particularly supervisory training, must be directed not merely at disseminating knowledge but rather at telling participants what management wants done and how to do it. After objectives are defined, content is generally easy to decide on. The last step should be the determination of method, which is too often used as the first step.

Today there is tremendous choice in method of instruction. In past years a trainer could lecture or discuss; today there are audio-visual aids of every description—films, slides, filmstrips, and TV programs, for example. In addition, there are large numbers of programmed instruction texts, management games, etc.

Whom and When to Train

The answer to this seems simple. Obviously, we train those who have a defined need for training. Thus we would train:

1. The new employee—orientation to the company
2. The new employee—on-the-job training
3. The experienced worker—when needed
4. Any employee transferring to another job
5. The group—toolbox meetings
6. Other employees as the need arises

What to Provide Training In

The answer to this also seems simple. The content of a training program is what management wants employees to know. More specifically, the training should teach them how to carry out the responsibilities that management has assigned.

Actually, content, or *what* to train, is decided as objectives are identified. If objectives are stated in terms of what a participant should be able to do upon completion of the training, selection of content is simple. For instance, if one objective is that the participant should be able to make a job safety analysis of each job, one segment of the training will consist of actual training in using the tool.

THE INFLUENCE OF SUPERVISION

Practically all our motivational attempts get to employees through their supervisors. The supervisor is, among other things, a funnel, directing all our material and information to the employee. He or she also directs or carries out the vast majority of training. Everything that motivates employees is applied by the supervisor, and obviously this role is crucial. The supervisor, of course, must be motivated and must share the company's goal if we are ever to achieve our purposes. Chapter 12 discusses in detail the motivation of the supervisor.

THE INFLUENCE OF SPECIAL-EMPHASIS PROGRAMS

A special-emphasis program is not the same thing as a gimmick or a contest. It is a coordinated campaign aimed at a defined problem. A well-run and well-devised special-emphasis program can exert an important influence on employees.

One example of a good program is the "Knowing's Not Enough" program put together a number of years ago by the United States Steel Company. This program was conceived under the assumption that employees know what to do—they just do not do it often enough. The campaign materials included an excellent film, literature, posters, and a recall symbol. Properly used, it has been extremely effective. Another example of a special-emphasis program is the "Everywhere—All the Time" campaign devised by the Allis-Chalmers Company and later adopted by the National Safety Council. Campaign content was similar to that of the "Knowing's Not Enough" program, and the defined problem in this case was off-the-job injuries.

One final special-emphasis program is the NO STRAIN campaign, put together by Industrial Indemnity Company, which is aimed at the defined problem of the industrial back strain, the biggest single accident type. This campaign does not require a film, and it comes packaged and ready for use in a kit that contains all needed materials (posters, banners, visual teaching aids, wallet cards, literature for employees and supervisors, etc.). It includes a manager's guide which gives detailed instruction on how to run the campaign. (See Appendix C for details.)

Special-emphasis programs are effective because (1) they are directed at defined problems, (2) they utilize a recall symbol which triggers response over and over, and (3) they contain enough elements to convince employees that management is really interested in their safety. If management were not, would it have gone to all that trouble?

Safety Media

Perhaps the role of safety media has been downgraded too much in this book. Media are not without value. On the contrary, they are extremely valuable—but only when heard by receptive ears and seen by receptive eyes. The individual must be motivated to utilize media properly—media do not motivate by themselves. This principle has been proved in several research studies showing that the value of safety media is dependent not on the quality of the media itself but on the recipient's judgment of the company's safety activities.

If employees believe that management is genuinely interested in them and in their safety, the media (regardless of type) will be effective. Conversely, if employees believe that management is not interested in their safety, the media used will be ineffective.

The effectiveness of media, then (poster, literature, brochures, films, etc.), is felt only after we have a somewhat motivated work force. When employees know that management wants safety, they will be ready to perform with safety in mind, and at this point (and not until then) effective use can be made of safety media.

We can create a little temporary interest and perhaps create some group pressures through gimmicks and contests. This will result in a temporarily improved record, but it takes more to achieve a permanently good accident record. It takes genuine interest from the top, expressed in the form of management direction of safety.

MOTIVATORS AND DISSATISFIERS

A number of studies on industrial motivation have been made in recent years. All of them substantiated the notion that the factors that motivate people are separate and distinct from the factors that dissatisfy people. The studies found the following to be "motivators" and "dissatisfiers":

Motivators	*Dissatisfiers*
Achievement	Company policies
Recognition	Supervision
Quality of work	Working conditions
Responsibility	Salary
Advancement	Relationships
	Status

The studies indicated that the factors labeled as possible dissatisfiers should be brought up to a level of "content." When this level is reached, additional improvement does not really serve us. The employees are satisfied, but that does not mean that they are motivated. The next step is to work on the motivators.

Application of Theory

How can we utilize our knowledge of the function of motivators? That is, how can we use it to motivate? We might:

	Motivator
1. Increase the accountability of individuals for their own work	Recognition and responsibility
2. Remove constraints and keep accountability	Achievement and responsibility
3. Grant more authority or job freedom	Recognition and responsibility
4. Give new, more difficult tasks	Growth
5. Let the worker in on company information	Recognition

Now, how can we use such information in safety? Looking at the list of dissatisfiers above, we find that some of the items apply very directly to safety:

Company policy. Often our too stringent policies on safety, together with a lack of direction in other areas, become a strong dissatisfier.

Supervision. The quality of supervision surely affects safety motivation as well as production levels.

Working conditions. Unsafe conditions allowed to exist by management lead to strong dissatisfaction with regard to safety. Employees asked to do something in safety will react unfavorably when there is evidence that management is not doing its part.

The above are some examples of dissatisfiers; most workers could add to the list.

The first order of business in safety should be to bring these dissatisfiers up to level of content. Once this is accomplished, we can start on the motivators. We might:

	Motivator
1. Involve employees in setting safety standards	Responsibility and recognition
2. Provide a self-monitoring system	Responsibility and achievement
3. Make improving safety procedures a part of every worker's job	Responsibility and achievement
4. Share reports on safety	Recognition
5. Recognize safe procedures	Recognition

For instance, take item 1 above. Traditionally, safety rules have been developed by the safety department or by management and then handed down from the top as though they were stone tablets. The employees had nothing to say about them. Their natural tendency was to resist such rules rather than obey them. All people value their own decisions; hence employees are far more likely to accept and live by safety rules if they play a part in their development. Then motivation to follow the rules comes from within the group rather than from outside it, from management.

Robert Burns, formerly with General Electric Company, puts it this way:

> In a participative safety program, the supervisor leads the group in developing its rules, while the safety director is the expert whom the group consults for professional assistance. Both men should encourage members of the group to express their opinions and should ask numerous questions to be certain that an issue has been thoroughly discussed. With their help, the group will develop safety standards that meet the needs of the department.
>
> Once workers decide what rules of safety are applicable to their department, they will actually enforce the rules themselves. The few employees who insist on violating the rules will be persuaded to conform to safety standards under threat of ostracism. Since the need to be accepted by the group is a powerful motivator, the violator will usually adopt the safe methods.
>
> The participative approach to establishing safety rules can also spur an employee's sense of individual commitment to the overall objectives of the department. His role in formulating the rules he must obey reinforces his self-respect and affirms his value to the department. The satisfaction he gains will stimulate him to contribute to his department's efforts in other areas besides safety.[1]

REFERENCES

Burns, R. L.: "Why Not Try Participative Safety?" *Supervisory Management,* March 1969.

Campbell, R. K., and **D. C. Petersen:** "No Strain Campaign," Industrial Indemnity Co., San Francisco, 1969.

Herzberg, F.: "One More Time: How Do You Motivate Employees?" *Harvard Business Review,* January–February 1968.

Hughes, C.: "Making Safety a Meaningful Part of Work, *Journal of the ASSE,* June 1967.

Likert, Rensis: *New Patterns of Management,* McGraw-Hill Book Company, New York, 1961.

Myers, N. S.: "Who Are Your Motivated Workers?" *Harvard Business Review,* January–February 1964.

Petersen, D. C.: "Six Management Tools for Industrial Safety," *Safety Maintenance,* July 1966.

Weaver, D. A.: "Strengthening Supervisory Skills," Employers Insurance of Wausau, Wausau, Wis.

———: "When You're Speaking of Safety," Employers Insurance of Wausau, Wausau, Wis., 1963.

[1]R. L. Burns, "Why Not Try Participative Safety?" *Supervisory Management,* March 1969.

PART FIVE

ADDITIONAL SAFETY TECHNIQUES

The safety function goes beyond mere safety of employees on the job. The function includes (1) employee on-the-job safety, (2) employee off-the-job safety, (3) nonemployee safety (liability), (4) safety in the company fleet vehicles, (5) property damage, (6) product safety, and (7) fire losses.

Obviously, not all safety professionals have assigned responsibilities in all these areas. Someone should, however. If someone else in your organization has these responsibilities, then you can overlook them. More often than not, however, no one has defined responsibilities except in item 1 above. If this is the case, then chances are that the safety professional has an implied responsibility in some of the other areas.

Two of the largest potential loss sources in most companies are vehicle accidents and product losses. Controls for both these areas are discussed in this part.

We shall also look at some excellent new measurement tools and appraisal tools and at some new management tools for unearthing causes of loss and for controlling loss.

14

TRACING SYSTEMS:TOR

One way to assist line managers in determining underlying causes of events is to provide them with a system that forces them to look at possible contributing factors.

In a tracing system the investigator is initially asked to identify what appears to be a major cause of, or factor behind, an event. He or she selects this from a large list of possible causes. Following the description of each major cause are numbers which lead the investigator to other factors that are generally connected to the initial factor and thus might be contributing causes.

One of the best tracing systems is TOR—Technic of Operations Review. It was devised by D. A. Weaver and centers around the cause code shown in Figure 14-1. The supervisor's incident investigation report in Figure 14-2 was devised for use with TOR by Paul Mueller, of the Green Giant Company. A distinctive feature of TOR is that although it was devised to be used by an investigator to trace causes individually, it can also be used in management meetings as a very effective training tool. Probably the best explanation of TOR comes from its author and creator, D. A. Weaver:

> You have referred to "Technic of Operations Review" as a tracing technique, and I suppose it is, but I prefer to think of it in simpler terms. I prefer to present TOR analysis as simply a technique to create a sort of instant case study based on an actual incident.
>
> A case study usually involves a hypothetical set of facts about a hypothetical company, and discussion creates hypothetical insights. TOR analysis creates real insights into the real circumstances of a real organization. Reality begins with the fact that TOR analysis is applied to an incident, some untoward event, that really occurred in the organization.
>
> An accident or an injury is the clearest example of an incident, but the word incident includes all sorts of wasteful and inefficient occurrences. One man spoke of "300 gallons of product down the drain," a memorable incident since the product was beer. Another identified "$9,000 worth of scrap." Another described a costly delay and

TOR TECHNIC OF OPERATIONS REVIEW

1 COACHING

10 Unusual situation, failure to coach (new man, tool, equipment, process, material, etc.) 44, 24, 62

11 No instruction. No instruction available for particular situation 44, 22, 24, 80

12 Training not formulated or need not foreseen ... 24, 34, 86

13 Correction. Failure to correct or failure to see need to correct 42, 20, 30

14 Instruction inadequate. Instruction was attempted but result shows it didn't take 15, 16, 42

15 Supervisor failed to tell why 44, 24, 83

16 Supervisor failed to listen 11, 81

17

18

19

2 RESPONSIBILITY

20 Duties and tasks not clear 44, 34, 14, 53

21 Conflicting goals 80,

22 Responsibility, not clear or failure to accept ... 26, 14, 54, 82

23 Dual responsibility 47, 34, 13

24 Pressure of immediate tasks obscures full scope of responsibilities 36, 12, 51

25 Buck passing, responsibility not tied down 44, 26, 55, 60

26 Job descriptions inadequate 80, 86

27

28

29

3 AUTHORITY (Power to decide)

30 Bypassing, conflicting orders, too many bosses 44, 13

31 Decision too far above the problem 36, 83, 85

32 Authority inadequate to cope with the situation 81, 83

33 Decision exceeded authority 20, 26, 14

34 Decision evaded, problem dumped on the boss 36, 14, 85

35 Orders failed to produce desired result. Not clear, not understood, or not followed 40, 46, 13, 15

36 Subordinates fail to exercise their power to decide ... 26, 12, 83, 85

37

38

39

4 SUPERVISION

40 Morale. Tension, insecurity, lack of faith in the supervisor and the future of the job 15, 56, 64, 80

41 Conduct. Supervisor sets poor example 13, 84

42 Unsafe Acts. Failure to observe and correct 24, 11, 52

43 Rules. Failure to make necessary rules, or to publicize them. Inadequate follow-up and enforcement. Unfair enforcement or weak discipline 25, 36, 12, 52

44 Initiative. Failure to see problems and exert an influence on them 22, 34, 30

45 Honest error. Failure to act, or action turned out to be wrong 10, 12, 15, 81

46 Team spirit. Men are not pulling with the supervisor 40, 21, 56

47 Co-operation. Poor co-operation. Failure to plan for co-ordination 23, 25, 15, 66

48

49

FIGURE 14-1 TOR review cause code

5 DISORDER

51 Work Flow. Inefficient or hazardous layout, scheduling, arrangement, stacking, piling, routing, storing, etc. 41, 24, 31, 80

52 Conditions. Inefficient or unsafe due to faulty inspection, supervisory action, or maintenance 21, 32, 14, 86

53 Property loss. Accidental breakage or damage due to faulty procedure, inspection, supervision, or maintenance . 43, 20, 80

54 Clutter. Anything unnecessary in the work area. (Excess materials, defective tools and equipment, excess due to faulty work flow, etc.) 44, 36, 80

55 Lack. Absence of anything needed. (Proper tools, protective equipment, guards, fire equipment, bins, scrap barrels, janitorial service, etc.) 44, 36, 80

56 Voluntary compliance. Work group sees no advantage to themselves 40, 15, 41

57

58

59

6 OPERATIONAL

60 Job procedure. Awkward, unsafe, inefficient, poorly planned 44, 32

61 Work load. Pace too fast, too slow, or erratic 44, 51, 63

62 New procedure. New or unusual tasks or hazards not yet understood 43, 44

63 Short handed. High turnover or absenteeism 80, 40, 61

64 Unattractive jobs. Job conditions or rewards are not competitive 81, 46

65 Job placement. Hasty or improper job selection and placement 80, 86

66 Co-ordination. Departments inadvertently create problems for each other (production, maintenance, purchasing, personnel, sales, etc.) 45, 35, 13

67

68

69

7 PERSONAL TRAITS
(When accident occurs)

70 Physical condition — strength, agility, poor reaction, clumsy, etc. 44, 26, 65

71 Health — sick, tired, taking medicine 44, 24, 65

72 Impairment — amputee, vision, hearing, heart, diabetic, epileptic, hernia, etc. 44, 24, 65

73 Alcohol — (If definite facts are known) 80

74 Personality — excitable, lazy, goof-off, unhappy, easily distracted, impulsive, anxious, irritable, complacent, etc. 44, 13

75 Adjustment — aggressive, show off, stubborn, insolent, scorns advice and instruction, defies authority, antisocial, argues, timid, etc. 44, 13

76 Work habits — sloppy. Confusion and disorder in work area. Careless of tools, equipment and procedure 44, 13

77 Work assignment — unsuited for this particular individual 42, 65

78

79

8 MANAGEMENT

80 Policy. Failure to assert a management **will** prior to the situation at hand 24, 81, 83

81 Goals. Not clear, or not projected as an "action image" 83, 86

82 Accountability. Failure to measure or appraise results . 36

83 Span of attention. Too many irons in the fire. Inadequate delegation. Inadequate development of subordinates 12, 86

84 Performance appraisals. Inadequate or dwell excessively on short range performance 20, 65

85 Mistakes. Failure to support and encourage subordinates to exercise their power to decide 36

86 Staffing. Assign full or part-time responsibility for related functions 66

87

88

89

FIGURE 14-1 TOR review cause code (*continued*)

frantic search for a special truck shipment which all the while sat unbeknownst at the dock. Such incidents trigger TOR analysis and the search for insight and understanding into the operating errors which produce them.

Such incidents of greater or lesser significance occur in every business day, often hardly observed. Someone corrects the symptoms, but no one seeks the causes, the underlying operating errors; and things roll on until a costly snafu occurs. Then the boss gets into the act to find out who was to blame, and of course no one is.

SUPERVISOR'S INCIDENT INVESTIGATION REPORT

Employee (if involved) Dept. Clock No.

Incident date Reported

1. Describe the incident. Include location, witnesses, and circumstances surrounding incident. Try to identify the causal factors involved.

2. Subject causes to TOR analysis: state, trace, eliminate.

3. List factors for which you will initiate corrective action.

Factor: Action:

4. List factors which require feasible corrective action by others. Circle routing to their attention.

FIGURE 14-2

"Technic of Operations Review" thus begins with an incident. It begins with an incident, not a problem. The distinction is important, for TOR analysis is not a tool for problem solving. It is a tool to help "locate and define operating error." The distinction can be seen in a case in which a manager was advised that TOR analysis could not be applied to his "problem" of absenteeism. He thought a bit and then said, "Well, one day, 90 percent of my crew failed to show up. Now, do I have an incident?" Yes indeed he does, and TOR analysis exposed many of the underlying causes of his recurrent problem. Thus his managerial effort was directed to causes rather than to his problem of absenteeism which turned out to be only a symptom of his real problems.

Accidents and injuries are simply one kind of incident, a relatively rare and quite definable kind of incident. Except for accidents, it is difficult to define the word "incident," but it is easy to describe them and to find examples. For the moment, the important thing is that incidents are not problems. Incidents trigger TOR analysis. Its misuse as a problem-solving tool creates confusion and disappointment. Its purpose is to expose what the problems are, to expose operating errors.

To illustrate TOR analysis, picture an incident as a manager of Excel Devices Company first hears of it. A customer has received a palette load of devices as ordered but not one of the devices contained its essential electric motor. A man was dispatched with a box of motors to install them on the spot. Here's an irate customer, phone calls, travel expense, overtime, waste, and inefficiency. Instead of blaming someone, he sits down with his supervisors in a TOR session.

The incident is described and brief question-answer discussion reveals the facts of the case. Analysis of anything requires facts. So the first five minutes or so is devoted to facts, not blame.

Next, the manager directs his supervisors to the TOR form and directs them to select one number which they consider to be the direct proximate cause of the incident [see Figure 14-1]. He asks them to STATE IMMEDIATE CAUSE, the first step of TOR analysis. Discussion and disagreement ensue. The form must be learned, words and meanings hashed, and insights gained. This first step may absorb five or ten minutes (the manager should press his supervisors for consensus agreement at this point, and not let discussion degenerate into an aimless bull session). By pressing his demand for a decision, the group finally agreed on item 43 as the IMMEDIATE CAUSE.

This number, 43, is jotted at the top of a sheet of paper and the TRACE STEP begins as illustrated below. Do not discuss each item in depth. The TRACE STEP begins as a hasty overview. The manager should press for a prompt decision, a minute or two on each item. Cause the group to decide briskly whether each item did or did not contribute to the incident, whether it is "in" or "out." Keep insisting on that decision to control aimless bull sessioning. Trace numbers are listed vertically as they develop. In this illustration, explanatory notes indicate the discussion and thinking that took place.

43 Rules
25 Buck-passing
36
~~12~~
~~52~~
44
~~26~~
~~55~~
60
~~26~~
~~12~~
83
85
22
34
30
44
32

Each box should have received a stricker indicating incomplete assembly. Instead, one sticker had been attached to the whole palette load. [For trace numbers follow item 43 on the TOR form, Figure 14-1.] These are listed vertically at left.

ITEM 25. Buck-passing contributed because it was not clear who should observe the sticker: the forklift operator, the loading foreman, or the assembly foreman. (Since item 25 is "in," four more trace numbers result.)

ITEM 36 Authority was ill used, the supervisors decided, since this rule violation is widely overlooked. Add numbers 26, 12, 83, 85 to the list.

ITEM 12. "out," not a factor; hence crossed out. Note: Discussion of just four points has generated a long list looming ahead. Have faith; the TRACE STEP does end.

Press on, urging "in" or "out" decisions; you want insight and hasty overview, not in-depth discussion.

ITEM 52. "out." Conditions not at fault.

36

ITEM 44. "in." Initiative at fault. Add numbers 22, 34, 30 to the list.

ITEM 26. "out." Job descriptions were adequate.

ITEM 55. "out."

ITEM 60. "in." Assembly line frequently failed to receive needed parts thus producing incomplete devices at the end of the line. Add 44, 32 to the list.

ITEMS 26 and 12. "out" as previously decided.

ITEM 85. "in." Supervisors decide that mistakes were blamed, but inaction was overlooked. Add 36 to the list as shown on the TOR form.

Tracing is not yet finished, but it is more exciting to observe this process in a group than it is to read it, so let us stop. In practice, the TRACE STEP continues until it runs out in either of two ways. Usually, the "outs" overtake the "ins" and you come to the end. Or, sometimes, the final number on the list repeats the number at the top (in this case, item 43), and you have come full circle back to the beginning.

So far, we have illustrated the STATE step and have demonstrated the TRACE step, even though we leave it unfinished in the illustration. When the TRACE step is completed, two steps remain. To illustrate the next step, ELIMINATION, let me complete tracing without explanation. In brief, item 30 was accepted as contributing to the incident, and this in turn generated item 13 as "in." Thus from all the items assessed, we end up with the following list of operating errors which contributed to the incident, the following list of "in" items:

43
25
36
44
60
83
85
30
13

ELIMINATION begins by listing the "in" items, in this case nine of them. Management cannot correct nine things at once; hence the ELIMINATION step. Elimination reduces the list to discussing items in terms of importance and prevalence.

In this case, for example, items 13 and 30 were regarded as exceptions, not prevalent supervisory performance. Items 44, 36, and 25 were regarded for the moment as self-corrective as a result of the insights gained in the TOR session. The elimination meant discussion and insights in depth, which can be appreciated by noting the meanings attached to the numbers. The elimination slid into the final step, SEEK FEASIBLE CORRECTIVE ACTION.

43
60
83
85

SEEK FEASIBLE CORRECTIVE ACTION, the last step, focuses in this case on just four items instead of nine. The supervisors themselves proposed that the rule requiring a sticker on each incomplete device would be enforced.

They further proposed to put the stickers on the side from which the forklift driver approached. Thus he would be sure to see them.

The manager began an investigation to determine why parts frequently failed to appear at the assembly line in proper coordination. He was also left pondering items

83 and 85, which had stayed "in," revealing that his supervisors felt that mistakes were noted but failure to act was largely overlooked.

I believe that TOR helps to identify a problem by tracing symptoms back to underlying root causes. A problem correctly defined is usually half solved. TOR helps in identifying and in defining problems by searching for root causes, causes that lead eventually to accidents and numerous other kinds of management losses.

15

SYSTEMS SAFETY FOR THE HIGH-POTENTIAL ACCIDENT

In Chapter 2 we briefly discussed the high-potential accident. Principle 2 in that chapter stated that certain sets of circumstances can predictably produce severe injuries and that these potential severe injuries can be identified and thus controlled. Several situations were identified as having a high probability for severity. Unusual or nonroutine tasks, any nonproduction work, work carried out near high-energy sources, and certain construction situations were listed as examples.

In recent years many studies have been made of the high-potential accident. High-potential accidents are defined as those which did, or under similar circumstances (or slightly different ones) could, result in serious injury or damage. High-potential accidents can include the so-called "near misses" in which no injury occurs and incidents in which a person receives only a minor injury.

Much of the material currently being produced in this area seems to be aimed at helping to locate and identify the high-potential accident. For instance, of two similar injuries—say, pinched index fingers—one could be a minor injury with little potential seriousness (if the pinch came from a pair of pliers), and the other could be a high-potential accident (if the finger was nipped in the calender rolls of a paper machine). Most material has been aimed at systems and techniques to help management and supervisors define and decide what accidents are high-potential and thus worth further investigation and time.

Sophisticated systems have been developed to identify these high-potential accidents, and new forms are in use in some companies for reporting them to safety. Some companies utilize interviewing techniques to help discover some of the incidents that could be high-potential in nature.

Relatively less has been written about the control of exposures leading to high-potential accidents. Perhaps some connection and correlation can be made between the high-potential accident work under way and the field of systems safety.

DEFINITION OF SYSTEMS SAFETY

Systems safety is still a somewhat unknown discipline to the industrial safety professional. Although at times we feel that it is of little or no value to us, there are, no doubt, concepts in systems safety that could be usefully applied to industrial safety.

Industrial safety and systems safety start from a common base—the desire to save lives and property. Yet systems safety is oriented toward analysis and improvement of hardware (that is, systems), and industrial safety directs itself to people more and more each year. Both start from a common base, but they start with different fields of endeavor. Industrial safety strives primarily to control accidents to employees on the job. So far, systems safety has worked mainly in the area of product safety in the aerospace and automotive fields.

Industrial safety engineers operate in a fixed manufacturing situation. They work in the midst of hazards that have often been there for a long time, many of which are accepted by production as a necessary component of their way of operating. They must work within that framework, perhaps teaching the employees how to work around those hazards instead of removing them.

The first concern of the systems safety engineer is that a given system work as it has been designed to, that is, that the design be foolproof and that no one can possibly get hurt. For instance, in the case of a spacecraft, the systems safety engineer would be vitally concerned that the retro-rockets fire—on schedule—or all would be lost for the astronauts. Systems safety techniques, then, are the result of a need to eliminate any hardware malfunctions or mistakes in design that could have serious consequences. In many of these situations the only safeguard possible is to analyze or anticipate problems and then design in order to avoid them.

There are many methods of analysis in use in systems safety:

1. Gross-hazard analysis. This is done early in the design stage. It is the initial safety analysis, and it considers the overall system.

2. Classification of hazards. Types of hazards are identified and classified with regard to potential severity.

3. Failure modes and effects. The kinds of failures that could happen are examined, and their effects are predicted.

4. Hazard criticality ranking. The probability of different hazards' occurring is determined, and the hazards are ranked in order from most to least critical.

5. Fault-tree analysis. Fault-tree analysis traces the progression of hazards.

Other types of analysis made in systems safety consider (1) high-energy potentials, (2) catastrophe accidents only, and (3) maintenance considerations of system skills required for operation of a system.

FAULT-TREE ANALYSIS

Fault-tree analysis is one system of making a detailed analysis of failure or potential failure. It was successfully applied initially in aerospace and has been used in other fields. It seems likely that the technique will be used in industrial safety in the future.

The fault tree is actually a logic diagram that traces all the events that might have led to the undesired result being studied. A fault tree can be constructed for any event. First, an undesired event is selected. Then it is necessary to reason backward to visualize and identify all the ways in which it might have occurred. Each contributing factor or cause is then studied and analyzed to determine how *it* could possibly have happened. Such tracing of causes and factors can point to many different system failures that might not otherwise be noticed.

Say that the undesired event selected is a severe injury to the operator of a power press in an industrial plant. The fault tree (see Figure 15-1) for this event is made by listing those events which might have occurred, or which must have

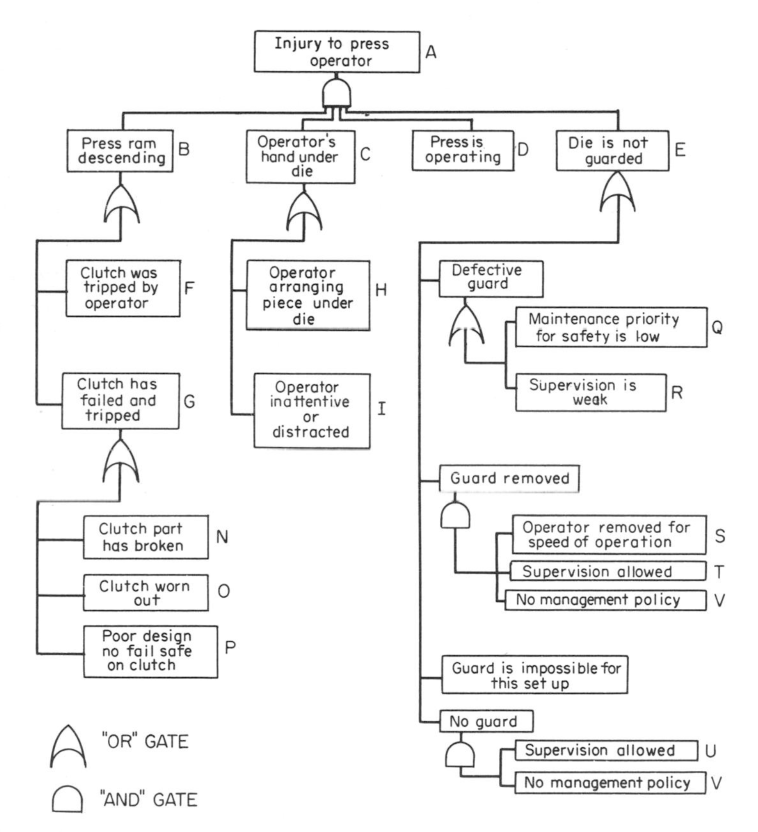

FIGURE 15-1 Fault-tree analysis

occurred, for the undesired event to happen. These events are connected by either AND or OR gates. An AND gate means that both events noted must be present for the event to occur. An OR gate means that either event alone can be responsible for the occurrence of the major event. Thus in Figure 15-1, events B, C, D, and E must all be present for the major undesired event A to occur.

However, either event F or event G alone could cause subevent B, and either H or I alone could cause subevent C. Event N, O, or P alone could cause subevent G.

This fault tree shows that for a severe injury to occur to the operator of the power press, four things must happen:

1. The press must be operating.
2. The ram must be descending.
3. The operator must have a hand under the die.
4. There must be an inoperative guard or no guard on the die.

Hence, prevention lies in the elimination of one of these four events.

Each of the four is then analyzed. For the press ram to be descending, the clutch must have been tripped by the operator, or the clutch must have failed, and the press repeated without being tripped. Any of these could happen if a clutch part had broken, the clutch was worn out, or the clutch had been poorly designed, with no fail-safe mechanism built into it.

The operator's hand might have been under the die if he or she had been working under it to arrange the piece or had been distracted and was paying poor attention to the work. The die could have been unguarded if the guard had been removed, if the guard had been in some way defective, if there was no guard, or if on this particular die and setup a guard was impractical or impossible.

Each of these subevents could then be further analyzed, which might lead to such conclusions as (Q) maintenance priorities do not include press die-guard construction, (R) press department supervision is not finding defective guards (is not looking for them), (S) press operators are removing guards, (T) supervision is not enforcing the use of guards, (U) supervision is not stopping work on a press when guards are absent, and (V) management is not setting and enforcing a policy of press guarding. (Notice that in this particular fault-tree analysis, human failures have been included.)

The above analysis is very simple; it was chosen merely to illustrate the general idea of fault-tree analysis and to indicate how it might possibly be used in industrial safety as well as in systems safety. You can no doubt visualize a complex analysis, for example, one that might be used in the aerospace or missile field.

Fault-Tree Analysis and High-Potential Accidents

Perhaps systems safety can be best used in industry in conjunction with the high-potential accident analysis approach. Using present-day systems of locating and defining high-potential accidents or high-potential accident exposures, we can periodically and regularly identify potential high losses.

After identifying the high-potential accident, the next step would be to make an estimate of the potential dollar loss that could reasonably be expected should the accident occur. For a high-potential accident that has already happened, this estimation would be relatively simple: The type of injury would be pictured, and the loss along the lines would be estimated as described in the costing process discussed in Chapter 4. Additional cost estimates can be made of any indirect costs, and property damage losses and a total estimated cost can be derived.

At some predetermined dollar loss level, a fault-tree analysis would be required. Since fault-tree analyses can be complex in some cases and hence time-consuming, this level would necessarily be high, perhaps $10,000 or $25,000. The analysis might be made by safety, or the system could be taught to line supervision, who would then make the analysis.

Both high-potential accident analysis and systems safety are proved techniques, but both are practically unused in industrial safety.

REFERENCES

Allison, W. W.: "High Potential Accident Analysis," *Journal of the ASSE,* July 1965.

Frazier, Dana: "A Critique of the High Potential Method," *Aerospace Newsletter,* National Safety Council, Chicago, 1966.

MacKenzie, E. D.: "On Stage for System Safety," *Journal of the ASSE,* October 1968.

Recht, J. L.: "Systems Safety Analysis: The Fault Tree." *National Safety News,* April 1966.

16

USING THE COMPUTER

In recent years most large organizations have begun to utilize the computer in their injury record keeping. Most companies use an approach to this based primarily on the American National Standards Institute's Standard Z16.2. Most insurance carriers also have computerized injury records that provide their insureds with a loss run of some type to assist them in their loss-control efforts. Insurance companies also usually gear up their programs on the basis of the Z16.2 standard. This standard includes a general but comprehensive list of items within numbered sequences that can be used in the development of a record analysis system. Utilizing this standard as a starting point, companies then develop their own systems of analyzing their accidents and injuries in categories such as nature of injury, accident type, part of body injured, hazardous condition, source of injury, and unsafe act.

A large number of items are usually listed in each of the categories, allowing the reporter to choose items that come closest to describing what happened in connection with the accident or incident being reported on. Obviously, then, one factor determining the accuracy of input is the initial identification of items which will be on this list. The code shown in Figure 16-1 is an example of a basic input document; it was adapted from the report of a supervisor working for a large contractor. Figure 16-2 shows the input document of a large insurance company.

WEAKNESSES IN SYSTEMS

There are some inherent weaknesses in these systems in terms of current thinking about accident causation. The entire Z16.2 approach to analysis is based on getting a detailed description of the circumstance surrounding an accident, rather than on identifying the contributing causes. It is in actuality more a reflection of the domino theory of accident causation than a reflection of the multiple-causation theory. Thus, utilizing the Z16.2 approaches, many computerized analysis programs end up giving us a great deal of easily accessible but not very useful information and little insight into why accidents are occurring. Figure 16-3 shows a printout utilizing these approaches.

1. Report number ______
2. Date of injury ______
3. Job location ______

4. INJURED EMPLOYEE (Please print) LAST NAME / FIRST / INITIAL	BADGE NO.	JOB NO.	TIME OF INJURY HOUR	MIN.	AM/PM

5. CRAFT
A. Blrmkr. B. Carp. C. Cm. Msn. D. Elec. E. Insul. F. IW G. Lab. H. MW I. OE J. Paint. K. PF L. Surv. M. Tmstr. N. Other

6. AGE GROUP
A. 18–20 B. 21–25 C. 26–30 D. 31–35 E. 36–40 F. 41–45 G. 46–50 H. 51–55 I. 56–60 J. 61–65 K. 66–70 L. 71 +

SEX
M
F

7. HOW IT HAPPENED: ______

8. | DR CASE | LOST TIME | FATAL | WITNESS ______ EMPLOYEE ______

9. NATURE OF INJURY
A. Abrasion, scratch
B. Amputation
C. Burn
D. Concussion
E. Conjunctivitis (eye irritation)
F. Contusion, crushing, bruise
G. Cut, laceration, puncture
H. Dermatitis, rash
I. Electric shock, electrocution
J. Flashburn (eye)
K. Foreign body
L. Fracture
M. Heatstroke, exhaustion
N. Hernia, rupture
O. Inflammation, irritation
P. Respiratory problems
Q. Sprains, strains, dislocations
R. Multiple injuries
S. Injury or disease (NOC)
T. Nonoccupational claim (record purposes only)

10. PART OF BODY
A. Head (not face, eyes, internal ears)
B. Ears internal (including hearing)
C. Eyes
D. Face (nose, mouth, lips, teeth, etc.)
E. Neck
F. Arm(s) (including wrist)
G. Hand(s) and or finger(s)
H. Abdomen (including internal organs)
I. Back (including muscles, spine, spinal cord)
J. Chest (including ribs, internal organs)
K. Hips (incl. pelvis, buttocks, internal organs)
L. Shoulders
M. Leg(s) (including ankle)
N. Foot and/or toe(s)
O. Multiple body parts
P. Circulatory system (heart, blood, vessels)
Q. Digestive system
R. Respiratory system
S. Body part (NOC)

11. SOURCE OF INJURY
A. Burning or welding operation
B. Hand tools
C. Power hand tools (except grinder)
D. Grinder (hand or pedestal)
E. Material, equipment being handled, worked
F. Body motion (not overexertion)
G. Ladder
H. Scaffold
I. Object improperly placed, located
J. Work surface, floor, ground, stair
K. Falling, flying, moving object
L. Building, structure, installed equipment, etc.
M. Rigging operation
N. Construction equipment, vehicle
O. Electrical equipment
P. Irritants (chemicals, dusts, fumes, etc.)
Q. Insects, plants, (wasp sting, poison oak, etc.)
R. Fire
S. Wind
T. Source of injury (NOC)

12. ACCIDENT TYPE
A. Struck against
B. Struck by
C. Fall from elevation
D. Fall on same level
E. Caught in, under, between
F. Overexertion (lifting, throwing, pulling, pushing)
G. Contact with electric current
H. Contact with temperature extremes
I. Contact with chemicals, dust, smoke
J. Accident type (NOC)

13. UNSAFE CONDITION
A. No unsafe condition
B. Defect of source of injury (dull, rough, sharp, slippery, broken, etc.)
C. Improper or inadequate clothing
D. Unsafe method or procedure
E. Unsafe placement or storage
F. Inadequate guarding
G. Inadequate illumination
H. Inadequate ventilation
I. Unsafe condition (NOC)

14. UNSAFE ACT
A. No unsafe act
B. Working on moving, energized, pressurized equipment
C. Failure to wear protective equipment (hard hat, safety glasses, rubber goods, foot guards, etc.)
D. Operating without authority
E. Failure to secure or warn
F. Horseplay
G. Improper use of equipment
H. Taking an unsafe position or posture
I. Inattention to footing or surroundings
J. Making safety devices inoperative
K. Undue haste
L. Driving error by equipment/vehicle operator
M. Unsafe loading, placing, mixing, etc.
N. Using unsafe equipment
O. Unsafe act (NOC)

FIGURE 16-1 Supervisor's report and injury analysis

As a result of our Z16.2 type of input, the computer has provided an analysis that shows a frequency of eight accidents caused by the employee being struck by a tipping, sliding, or rolling object, for a cost of $11,726. What are the real causes of these eight injuries? We have no idea. Where did they occur? We cannot tell from this analysis. What should we do about them? There is no way of knowing. A Z16.2 analysis does not deal with contributing causes and thus does not help us to direct our efforts. Perhaps the best information we get from this type of analysis is a breakdown of where accidents are happening (location, department, etc.), which is not a function of Z16.2.

INPUT DOCUMENTS

Thus, while computers are extremely valuable to us in assembling and analyzing data and information, the amount of assistance they provide is entirely

LOSS DESCRIPTION RECORD - SPECIAL ACCOUNTS

BASIC INFORMATION

1-2	4-11	12-17			18-25	26-37	38	39-40	41-46
Card	Policy number	Eff. Date			Claim Number	Claimant's Name	Sex	Age	Dept., Div. or Location
		MO.	DAY	YR.					
A4									

BEGIN NARRATIVE DESCRIPTION IN COLUMN 56

47-49	50-55			Description of Accident or Injury Columns 56-80					
State (ALPHA)	Accident Date			Acc. Type	Source of Inj.	Part of Body	Nature of Inj.	Area or Department	
	MO.	DAY	YR.	56-57	58-59	60-61	62-63	64-68	

NUMERIC CODING COLUMNS 56-68 ONLY

DESCRIPTION CONTINUED Use only when absolutely necessary. Recheck Basic description first before completing.

1-2	4-11	12-17			18-25	26-37	38	39-40	41-46
Card	Policy number	Eff. Date			Claim Number	Claimant's Name	Sex	Age	Dept., Div. or Location
		MO.	DAY	YR.					
B4	(COLUMNS 4 THROUGH 55 SAME AS ABOVE)								

47-49	50-55			56-80
State (ALPHA)	Accident Date			Description of Accident or Injury
	MO.	DAY	YR.	(LEAVE A BLANK SPACE BETWEEN WORDS, INITITALS OR SYMBOLS, USE UNDERSTANDABLE ABBREVIATIONS)

SUGGESTED ABBREVIATIONS

- **ABDOM** - ABDOMEN, ABDOMINAL
- **ADJ** - ADJUSTED, ADJUSTMENT
- **AMP** - AMPUTATED, AMPUTATION
- **BLDG.** - BUILDING
- **BRKN** - BROKEN
- **CONC** - CONCUSSION
- **ELEV** - ELEVATOR
- **EMPL** - EMPLOYEE
- **EXPL** EXPLOSION, EXPLODED
- **FLR** - FLOOR
- **FRACT** - FRACTURE, FRACTURED
- **FRT** - FRONT
- **FT** - FOOT, FEET
- **FTL** - FATAL
- **GRND** - GROUND
- **INF** - INFECTED, INFECTION
- **INJ** - INJURY, INJURED
- **LDR** - LADDER
- **L** - LEFT
- **MACH** - MACHINE
- **MULT** - MULTIPLE
- **OBJ** - OBJECT
- **OCC** - OCCUPATION
- **PASS** - PASSENGER
- **PEDEST** - PEDESTRIAN
- **PP** - PUNCH OR POWER PRESS
- **PROD** - PRODUCT
- **RR** - REAR
- **RUPT** - RUPTURED
- **R** - RIGHT
- **SCAFF** - SCAFFOLD
- **SDWK** - SIDEWALK
- **SHLDR** - SHOULDER
- **SLP** - SLIP, SLIPPED
- **SPRN** - SPRAIN, SPRAINED
- **STRK** - STRIKE, STRUCK
- **THRU** - THROUGH
- **TRK** - TRUCK
- **UNGD** - UNGUARDED

FIGURE 16-2 Input document, insurance company (*front*)

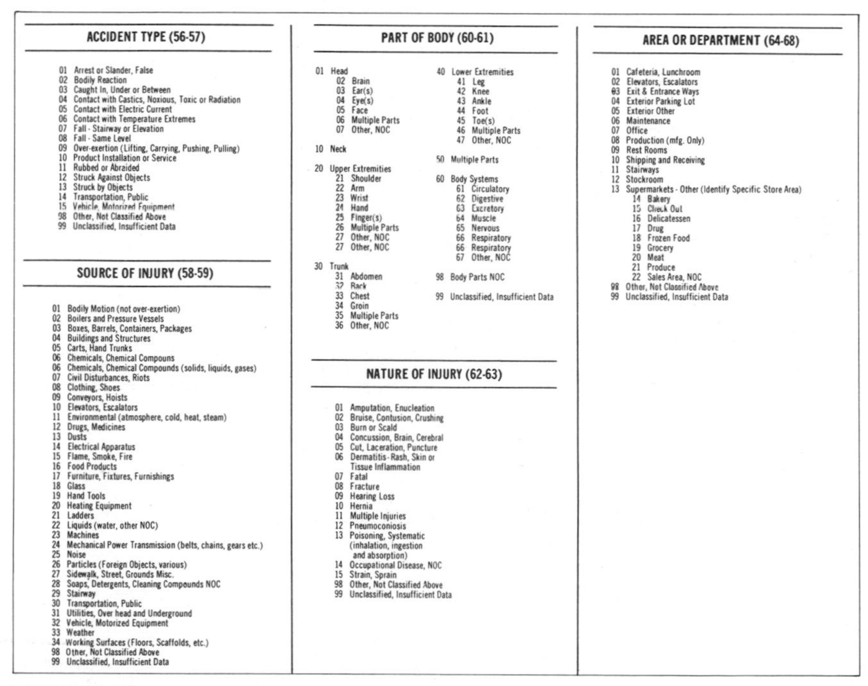

ACCIDENT TYPE (56-57)

01 Arrest or Slander, False
02 Bodily Reaction
03 Caught In, Under or Between
04 Contact with Castics, Noxious, Toxic or Radiation
05 Contact with Electric Current
06 Contact with Temperature Extremes
07 Fall - Stairway or Elevation
08 Fall - Same Level
09 Over-exertion (Lifting, Carrying, Pushing, Pulling)
10 Product Installation or Service
11 Rubbed or Abraided
12 Struck Against Objects
13 Struck by Objects
14 Transportation, Public
15 Vehicle, Motorized Equipment
98 Other, Not Classified Above
99 Unclassified, Insufficient Data

SOURCE OF INJURY (58-59)

01 Bodily Motion (not over-exertion)
02 Boilers and Pressure Vessels
03 Boxes, Barrels, Containers, Packages
04 Buildings and Structures
05 Carts, Hand Trunks
06 Chemicals, Chemical Compouns
06 Chemicals, Chemical Compounds (solids, liquids, gases)
07 Civil Disturbances, Riots
08 Clothing, Shoes
09 Conveyors, Hoists
10 Elevators, Escalators
11 Environmental (atmosphere, cold, heat, steam)
12 Drugs, Medicines
13 Dusts
14 Electrical Apparatus
15 Flame, Smoke, Fire
16 Food Products
17 Furniture, Fixtures, Furnishings
18 Glass
19 Hand Tools
20 Heating Equipment
21 Ladders
22 Liquids (water, other NOC)
23 Machines
24 Mechanical Power Transmission (belts, chains, gears etc.)
25 Noise
26 Particles (Foreign Objects, various)
27 Sidewalk, Street, Grounds Misc.
28 Soaps, Detergents, Cleaning Compounds NOC
29 Stairway
30 Transportation, Public
31 Utilities, Over head and Underground
32 Vehicle, Motorized Equipment
33 Weather
34 Working Surfaces (Floors, Scaffolds, etc.)
98 Other, Not Classified Above
99 Unclassified, Insufficient Data

PART OF BODY (60-61)

01 Head
02 Brain
03 Ear(s)
04 Eye(s)
05 Face
06 Multiple Parts
07 Other, NOC

10 Neck

20 Upper Extremities
21 Shoulder
22 Arm
23 Wrist
24 Hand
25 Finger(s)
26 Multiple Parts
27 Other, NOC
27 Other, NOC

30 Trunk
31 Abdomen
32 Back
33 Chest
34 Groin
35 Multiple Parts
36 Other, NOC

40 Lower Extremities
41 Leg
42 Knee
43 Ankle
44 Foot
45 Toe(s)
46 Multiple Parts
47 Other, NOC

50 Multiple Parts

60 Body Systems
61 Circulatory
62 Digestive
63 Excretory
64 Muscle
65 Nervous
66 Respiratory
66 Respiratory
67 Other, NOC

98 Body Parts NOC

99 Unclassified, Insufficient Data

NATURE OF INJURY (62-63)

01 Amputation, Enucleation
02 Bruise, Contusion, Crushing
03 Burn or Scald
04 Concussion, Brain, Cerebral
05 Cut, Laceration, Puncture
06 Dermatitis - Rash, Skin or Tissue Inflammation
07 Fatal
08 Fracture
09 Hearing Loss
10 Hernia
11 Multiple Injuries
12 Pneumoconiosis
13 Poisoning, Systematic (inhalation, ingestion and absorption)
14 Occupational Disease, NOC
15 Strain, Sprain
98 Other, Not Classified Above
99 Unclassified, Insufficient Data

AREA OR DEPARTMENT (64-68)

01 Cafeteria, Lunchroom
02 Elevators, Escalators
03 Exit & Entrance Ways
04 Exterior Parking Lot
05 Exterior Other
06 Maintenance
07 Office
08 Production (mfg. Only)
09 Rest Rooms
10 Shipping and Receiving
11 Stairways
12 Stockroom
13 Supermarkets - Other (Identify Specific Store Area)
14 Bakery
15 Check Out
16 Delicatessen
17 Drug
18 Frozen Food
19 Grocery
20 Meat
21 Produce
22 Sales Area, NOC
98 Other, Not Classified Above
99 Unclassified, Insufficient Data

FIGURE 16-2 Input document, insurance company (*back*)

CAUSE CODE	CAUSE OF ACCIDENT	FREQUENCY		CLAIM COST		PREMIUM	LOSS RATIO
		NEW	YTD	NEW CLAIMS	YR TO DATE		
	10 WORKMEN'S COMPENSATION						
10 01	EMPL STRUCK-INJURED BY HAND TOOL OR MACHINE IN USE		1		46		
10 02	EMPL STRUCK-INJURED BY FALLING OR FLYING OBJECT		6		584		
10 03	EMPL STRUCK BY TIPPING SLIDING OR ROLLING OBJECT		8		11,726		
10 11	EMPL STRAIN-INJURY LIFTING		2		202		
10 13	EMPL STRAIN-INJURY PUSHING OR PULLING		2		325		
10 17	EMPL CUT-SCRAPE BY HAND TOOL-UTENSIL-NOT POWERED	1	1	15	15		
10 19	EMPL CUT-SCRAPE BY OBJECT BEING LIFTED-HANDLED		1		52		
10 21	EMPL FELL OR SLIPPED ON SAME LEVEL		3		1,618		
10 22	EMPL FELL OR SLIPPED FROM DIFFERENT LEVEL	1	7	2,831	22,184		
10 23	EMPL FELL OR SLIPPED SLIPPED-DID NOT FALL		1		42		
10 31	STRIKING AGAINST OR STEP ON OBJECT BEING HANDLED		1		65		
10 32	STRIKING AGAINST OR STEP ON STEPPING ON SHARP OBJ		2		78		
10 42	CAUGHT IN OR BETWEEN MECHANICAL APPARATUS		1		513		
10 43	CAUGHT IN OR BETWEEN OBJECT HANDLED-OTHER OBJECT		5		325		
10 80	EMPL INJURY FOREIGN BODY IN EYE		8		602		
10 91	EMPL INJURY VEHICLE ACCIDENT - MISCELLANEOUS		9		14,294		
	TOTAL	2	58	2,846	52,671	81,341	64.8%
	OPEN		4		32,412		
	30 AUTOMOBILE LIABILITY						
30 25	AUTO LIAB - INTERSECTION OUR UNIT STRAIGHT ACROSS		5		136		
30 31	AUTO LIAB - NON INTERSECTION SIDESWIPE COLLISION	1	8	15,000	30,859		
30 32	AUTO LIAB - NON INT HIT PARKED OR STNDING VEHICLE		3		188		
30 40	AUTO LIAB - REAR END OUR UNIT HIT OTHER VEHICLE	1	4	350	4,559		
30 41	AUTO LIAB - REAR END OTHER VEHICLE HIT OUR UNIT		5		1,093		
30 42	AUTO LIAB - BACKING BACKING OR ROLLING BACK		6		706		
30 61	AUTO LIAB - FIXED OBJECT FIXED OBJECT - OVERHEAD		7		2,792		
30 67	AUTO LIAB TRACTOR TRAILER UNIT JACKKNIFED		1		13		
30 96	AUTO LIAB OBJECT FROM TRUCK	1	2		1,289		
30 99	AUTO LIAB MISCELLANEOUS-UNCLASSIFIED	1	1	82	82		
	TOTAL	4	43	15,432	41,717	106,602	39.1%

CLAIM CAUSE ANALYSIS PAGE 1 — CONTROL DATE 01-01 — ANALYSIS DATE 07-10

FIGURE 16-3 Computer printout showing analysis

	CAUSE CODE		EMPLOYEE DRIVER CLAIMANT	CAUSE OF ACCIDENT INJURY TYPE	PRESENT VALUATION	O F
OPEN CLAIMS						
12-05-70M	10 91	EMP	BETTES T A	EMPL INJURY VEHICLE ACCIDENT - MISCELLANEOUS	3,500	*O
			DRIVER	TRAILER JACKKNIFED-TRUCK RAN INTO EMBANKMENT		
12-05-70T	10 91	EMP	BETTES T A	EMPL INJURY VEHICLE ACCIDENT - MISCELLANEOUS	6,150	*O
			DRIVER	TRAILER JACKKNIFED-TRUCK RAN INTO EMBANKMENT		
08-18-70	50 15	DVR	JONES N	TRANSIT CARGO DAMAGE BY LOADING OR UNLOADING	1,000	*O
			UNIV MACH CO-CLMT	DROPPED GENERATOR WHILE UNLOADING	250	D
12-11-70	50 02	DVR	TUGWELL D	TRANSIT CARGO COLLISION OF OUR UNIT	450	*O
			US GOVEN-CLM	DMG TO PLASTIC GLASS CANOPY ON TRACTOR TANKS	250	D
12-27-70	50 16	DVR	LOVIETTE M T	TRANSIT CARGO IMPROPER PACKING	555	*O
			BECHTEL CORP-CLM	WOODEN SKID BROKE-DAMAGED COMPRESSOR	250	O
				TOTAL OPEN CLAIMS	11,655	
CLAIMS FINALIZED SINCE LAST REPORT						
06-27-70M	10 91	EMP	MCGEE R L	EMPL INJURY VEHICLE ACCIDENT - MISCELLANEOUS	187	F
			TRUCK DRIVER	VEHICLE STRUCK-BACK-NECK INJURIES		
06-27-70T	10 31	EMP	MCGEE R L	EMPL INJURY VEHICLE ACCIDENT - MISCELLANEOUS	1,107	F
			TRUCK DRIVER	VEHICLE STRUCK-BACK-NECK INJURIES		
11-23-70	10 91	EMP	JOHNSON H O	EMPL INJURY VEHICLE ACCIDENT - MISCELLANEOUS	0	F
			TRUCK DRIVER	SLIPPED OFF TRUCK-WRENCHED BACK-CUT SHIN		
11-29-70	10 03	EMP	THOMAS E J	EMPL STRUCK BY TIPPING SLIDING OR ROLLING OBJECT	158	F
			TRUCK DRIVER	HIT ON CHIN BY CHEATER PIPE-CUT LOWER LIP		
03-23-70	30 99	DVR	SPURLOCK	AUTO LIAB MISCELLANEOUS-UNCLASSIFIED	82	F
			UNKNOWN CLMT	FACTS UNKNOWN-AUTO PROPERTY DMG		
09-08-70	30 31	DVR	COLLEY T A	AUTO LIAB - NON INTERSECTION SIDESWIPE COLLISION	0	F
			JOHNSON R JR-CLMT	CLMT ATTEMPTED TO PASS-HIT DRIVERS VEHICLE		
09-09-70	30 32	DVR	BELL H	AUTO LIAB - NON INT HIT PARKED OR STNDING VEHICLE	0	F
			MEM FUNERAL HOME-CLMT	AMBULANCE KNOCKED INTO SIDE OF DVR VEH		
12-09-70	30 25	DVR	LITTLE R E JR	AUTO LIAB - INTERSECTION OUR UNIT STRAIGHT ACROSS	136	F
			PEL TRKG CO-CLMT	DR CROSSING INTERSECTION-STRUCK BY CLMT		
06-25-70	50 02	DVR	PAWKETT G	TRANSIT CARGO COLLISION OF OUR UNIT	9,761	F
			SIMMONS STEEL-CLMT	HIT TREE LIMB WITH LOAD-DMG CAB ON CRANE	250	D
11-18-70	50 02	DVR	COLLEY T A	TRANSIT CARGO COLLISION OF OUR UNIT	3,730	F
			WESTERN COM-CLMT	CLMT ALLEGES DAMAGES TO TOP OF OIL WELL UNIT	250	D
12-05-70	50 01	DVR	BARNES K D	TRANSIT CARGO COLLISION OF CARRIERS UNIT	2,500	F
			PAN AM PETRO CLMT	SCRUBBERS HIT OVERHEAD STRUCTURES ON BRIDGE	250	D
12-11-70	50 02	DVR	MASON R D	TRANSIT CARGO COLLISION OF OUR UNIT	132	F
			US GOVEN-CLM	DMG TO FLASTIC GLASS CANOPY ON TRACTOR TANKS	250	D
				LISTED FINAL CLAIMS	17,793	
				UNLISTED FINAL CLAIMS	12,088	
				TOTAL OPEN AND FINAL	41,536	
				TOTAL DEDUCTIBLES	2,250	

CLAIM COST CONTROL PAGE 1 CONTROL DATE 01-01- ANALYSIS DATE 07-10

FIGURE 16-4 Loss run

dependent on what we put in and on how we put it in. Figure 16-3 shows a printout that is a summary sheet. While most firms and insurance companies are able to provide such summary sheets, it is more common for them to provide simple loss runs, as shown in Figure 16-4.

This type of printout is, of course, even less useful in terms of directing our efforts in safety. It does, however, provide the kind of information that management often likes, for it shows who was injured, tells a little about how they were injured, and gives some dollar information. Two other typical analyses are shown in Figures 16-5 and 16-6. Figure 16-5 is a breakdown by job classification. This information is perhaps a little more usable, allowing us at least to zero in on types of jobs in which unusual frequencies or severities of accidents are being experienced.

Figure 16-6 breaks down the information by seriousness of accident or, rather, by cost of accident. It becomes obvious that there are some real problems connected with computer printouts dealing with dollars. In any claim involving a large amount, there is always a long delay before anyone has much of an idea as to the exact amount of the claim. In Figure 16-6, for instance, most of the information is based on "open" claims, which means that the dollar figures are only estimates of what the real costs will be. These estimates are often a far cry from the actual dollar figures, and for the most part they are almost unusable in our safety efforts.

Obviously, it is fairly easy to generate almost any kind of document you want if the information is already in the computer. And, as is also obvious, if the wrong information is in the computer, it is impossible to get usable information out. The systems which contain only Z16.2 information, and which therefore are unable to determine trends of causes, are a perfect example of this.

Another good example is the approach chosen years ago by some large insurance carriers to lighten the load of information to be coded and put into the computer. Only claims expected to cost over a certain amount are looked at and coded for insertion into the computer; all others are ignored. While this is perhaps fine from the standpoint of time saved in the company's home office, it certainly has an adverse effect on the usability of the information that comes out. One carrier that adopted this approach initially has, on the basis of the information in its computer system, been publishing data for a number of years on accident "causes" by type of industry, needed controls, etc. The inference of these data sheets, or technical guides, is that the information is based upon a comprehensive analysis of all accidents. It is not; only a few selected accidents are analyzed—those few that a worker in the claims department thought, at first glance, would cost over a preselected dollar figure.

It seems that analyses of accident causes and breakdowns for different industries coming from insurance carriers are somewhat suspect. Some insurance companies arbitrarily select a level of severity (cost) at which to begin collecting data, ignoring the rest. While there is nothing innately wrong with this, the companies usually do not tell us what their cutoff point is. Thus we

SUMMARY OF ANALYSIS BY JOB CLASSIFICATION
ANALYSIS PERIOD 10-01-69 TO 10-01-70
FOR COVERAGE 10 AND 11 WORKMEN'S COMPENSATION

PAGE NO. 022

	ACTUAL VALUES					COMPUTED VALUES BASED ON PROPORTIONAL DISTRIBUTION OF UNKNOWN				
	NO. CLAIMS	% TOTAL CLAIMS	VALUE	% TOTAL VALUE	AVERAGE VALUE	NO. CLAIMS	% TOTAL CLAIMS	VALUE	% TOTAL VALUE	AVERAGE VALUE
	0		$0		$0					
OFFICE MANAGER (CONST.)	1	.60	$133	.27	$133	1	.60	$133	.27	$133
DIST. WELDER	2	1.19	$43	.09	$21	2	1.19	$43	.09	$21
WELDER	2	1.19	$61	.12	$30	2	1.19	$61	.12	$30
WELDER 4 YEAR	1	.60	$10	.02	$10	1	.60	$10	.02	$10
STATION ATTENDANT	1	.60	$209	.42	$209	1	.60	$209	.42	$209
MAIN OPERATOR	8	4.76	$10,637	21.29	$1,329	8	4.76	$10,637	21.29	$1,329
ASST. MAIN OPERATOR	5	2.98	$609	1.22	$121	5	2.98	$609	1.22	$121
EQUIP. OPERATOR	5	2.98	$253	.51	$50	5	2.98	$253	.51	$50
FIELD BOOSTER ST. OP.	3	1.79	$143	.29	$47	3	1.79	$143	.29	$47
MAINT. MAN - REPAIRMAN	43	25.60	$7,020	14.05	$163	43	25.60	$7,020	14.05	$163
MAINT. MAN	31	18.45	$12,065	24.15	$389	31	18.45	$12,065	24.15	$389
REPAIRMAN,ROUSTABOUT,UTIL.HELP	5	2.98	$60	.12	$12	5	2.98	$60	.12	$12
LABORER	31	18.45	$3,213	6.43	$103	31	18.45	$3,213	6.43	$103
COR. CON. TECH. 1	1	.60	$11,000	22.02	$11,000	1	.60	$11,000	22.02	$11,000
UTILITY MAN #1	3	1.79	$182	.36	$60	3	1.79	$182	.36	$60
HANDLER	1	.60	$10	.02	$10	1	.60	$10	.02	$10
WELDER 3 YEAR	1	.60	$14	.03	$14	1	.60	$14	.03	$14
MACHINIST SR.	1	.60	$2,440	4.88	$2,440	1	.60	$2,440	4.88	$2,440

FIGURE 16-5 Analysis by job classification

CLAIM ANALYSIS BY SIZE RANGE — EXCLUDES DEDUCTIBLE

EXPERIENCE BY SIZE RANGE	EXPERIENCE PERIOD ONE NUMBER CLAIMS	% OF TOTAL	01-01-69 01-01-70 AMOUNT CLAIMS	% OF TOTAL	EXPERIENCE PERIOD TWO NUMBER CLAIMS	% OF TOTAL	01-01-70 01-01-71 AMOUNT CLAIMS	% OF TOTAL	EXPERIENCE PERIOD THREE NUMBER CLAIMS	% OF TOTAL	01-01-71 07-10-71 AMOUNT CLAIMS	% OF TOTAL
ZERO VALUE CLAIMS	724	10.6%		.0%	686	11.8%		.0%	133 2-OPEN	6.2%		.0%
1 TO 100	4,708 11-OPEN	69.3%	144,209 350-OPEN	5.0%	3,920 48-OPEN	67.6%	125,182 1,620-OPEN	5.4%	1,626 1,184-OPEN	76.0%	51,466 33,780-OPEN	7.5%
101 TO 250	523 2-OPEN	7.7%	82,815 383-OPEN	2.8%	462 5-OPEN	7.9%	72,287 1,016-OPEN	3.1%	138 30-OPEN	6.4%	21,502 4,827-OPEN	3.1%
251 TO 500	270 3-OPEN	3.9%	94,510 1,070-OPEN	3.2%	222 9-OPEN	3.8%	77,486 3,533-OPEN	3.3%	56 3-OPEN	2.6%	19,398 1,264-OPEN	2.8%
501 TO 1,000	186 4-OPEN	2.7%	131,537 3,120-OPEN	4.5%	134 8-OPEN	2.3%	95,103 6,038-OPEN	4.1%	19 6-OPEN	.8%	14,261 4,813-OPEN	2.0%
1,001 TO 2,500	181 22-OPEN	2.6%	286,257 44,674-OPEN	9.9%	205 65-OPEN	3.5%	342,202 130,304-OPEN	14.7%	118 101-OPEN	5.5%	235,196 208,844-OPEN	34.4%
2,501 TO 5,000	81 27-OPEN	1.1%	274,135 88,378-OPEN	9.5%	81 49-OPEN	1.4%	270,596 163,362-OPEN	11.6%	34 33-OPEN	1.5%	102,740 98,962-OPEN	15.0%
5,001 TO 10,000	60 28-OPEN	.8%	440,097 213,502-OPEN	15.2%	43 37-OPEN	.7%	315,810 274,826-OPEN	13.6%	5 5-OPEN	.2%	44,971 44,971-OPEN	6.5%
10,001 TO 25,000	41 31-OPEN	.6%	647,442 497,086-OPEN	22.5%	30 26-OPEN	.5%	472,033 417,720-OPEN	20.4%	8 8-OPEN	.3%	93,468 93,468-OPEN	13.6%
25,001 TO 50,000	9 9-OPEN	.1%	313,519 313,519-OPEN	10.9%	6 6-OPEN	.1%	219,732 219,732-OPEN	9.4%		.0%		.0%
OVER 50,001	5 5-OPEN	.0%	463,109 463,109-OPEN	16.0%	2 2-OPEN	.0%	323,809 323,809-OPEN	13.9%	1 1-OPEN	.0%	100,600 100,600-OPEN	14.7%
INCURRED TOTAL	6,788 142-OPEN	100.0%	2,877,630 1,625,191-OPEN	100.0%	5,791 255-OPEN	100.0%	2,314,240 1,541,960-OPEN	100.0%	2,138 1,373-OPEN	100.0%	683,602 591,529-OPEN	100.0%

ACCIDENT CAUSE ANALYSIS

FIGURE 16-6 Analysis by cost ranges

FORM DI-134 (face)

UNITED STATES DEPARTMENT OF INTERIOR
(FOR SAFETY MANAGEMENT USE ONLY)
SUPERVISOR'S REPORT OF ACCIDENT
Refer to DI-134A for Instructions, Definitions and Standard Coding Details

Field Report No.
DATE OF THIS REPORT

SECTION A. IDENTITY (Supervisor to Complete)

(1) ORGANIZATIONAL UNIT (Area, Region, etc.)
Last No. Here

(2) REPORTING STATION (Name & Address)
LAST NO. HERE

(3) STATE IN WHICH ACCIDENT OCCURRED

(4) Mo.____ Day____ Year____

(5) NEAREST HOUR OF ACCIDENT
____ a.m. ____ p.m.

(6) NAME (Employee, visitor or other involved)
Last ____ First ____
Soc. Sec. No.
Use separate form for each employee involved

(7) EMPLOYMENT STATUS (Circle one)
1 Permanent*
2 Y.O.C.
3 Temporary
4 Emergency
5 Contractor
6 Vista
7 Job Corpsman
8 Public (visitor)
9 Public (other)
0 Other (specify)
*Include Job Corps Staff

(8A) CSC OCCUPATIONAL CODE
Last No. Here
(8B) WORK ENVIRONMENT

(9) RESULT OF ACCIDENT (Circle one)
01 PI (personal injury)
02 PI with PF (fire)
03 PI with PM (motor veh.)
04 PI with PB (boat)
05 PI with PA (aircraft)
06 PI with PO (all other)
07 PF (fire)
08 PM (motor vehicle)
09 PB (boat)
10 PA (aircraft)
11 PO (all other property damage)

(10A) IDENTIFICATION OF PROPERTY DAMAGE (if any). (Give name, model, number, size, make, type, etc.)
Employee operated:

"Other" operated:

(10B) Make of Govt. Vehicle ____

(11) PROPERTY OWNERSHIP (Circle one)
0 No prop. involved
1 Interior owned
2 Other Federal
3 Contractor
4 Concession
5 Employee-owned on O.B.
6 Inter-agency (GSA) motor pool
7 Interior Leased
8 Privately owned
9 Other (Explain in No. 18)

(12A) AGE:
a. Of Employee (Year of birth) ____
(12B) b. Year of Mfg. ____

(13) IS TORT CLAIM EXPECTED? Circle one: Yes or No
If "Yes", has Tort Claim Officer been contacted? Circle one: Yes or No

SECTION B. MEDICAL (Supervisor to Complete)

(14) NATURE OF INJURY

(15) PART OF BODY INJURED

(16A) SEVERITY OF INJURY (Circle one)
0 No injury involved
1 First aid attention only
2 Medical attention only
3 Disabling injury (fatal)
4 Disabling injury (temporary)
5 Disabling injury (permanent)

(16B) Z16 COMPLIANCE
__Chargeable
__Not Chargeable

(16C) BEC FORMS PREPARED
Circle: Yes No

(17)
a. Leave date: Mo.____ Day____ Year____
b. Return date: Mo.____ Day____ Year____
c. Death date: Mo.____ Day____ Year____
d. No. days lost

SECTION C. STORY OF ACCIDENT (By Supervisor or Employee)

(18) NARRATIVE: INCLUDE: WHO, WHAT, WHEN, WHERE, AND HOW
Facts are important—fault finding not

NOTE: Condense Story Here

SECTION D. SUPERVISORY OPINION (Supervisor to Complete)

(19) I THINK ACCIDENT MIGHT HAVE BEEN PREVENTED IF:

Employee Fault Finding Adds Nothing to Management Improvement

(20) I SUGGEST THE FOLLOWING POLICY OR PROCEDURE CHANGE BY MGMT TO HELP PREVENT SIMILAR ACCIDENTS:

SIGNATURE AND TITLE OF REPORTING SUPERVISOR

FIGURE 16-7 Source document from the Department of the Interior

have no idea what is being analyzed. This, however, is probably not the biggest problem with such information.

The biggest problem is the validity and accuracy of the input. Accident reports are received in the claims departments of insurance companies, and then they must be coded for insertion into the computer system. There is only one place that this coding can be accurately done, and that is at the place of business where the accident happened. If coding is not done there (and it usually is not), it must be done at the insurance company. In some way, someone at the insurance company must translate the description given on the

ANALYSIS OF MANAGEMENT PROBLEM

FORM DI-134 *(back)* If the title you need is not in DI-134A, tell your safety officer so that it may be added. Information on this side represents a "team" effort by safety and other functional managers to assist the Line Officer to discover underlying causes of accidents and to plan for their ultimate correction. All items require an entry. Follow bureau instructions for analyzing the problems on this side.

SECTION E. LINE MANAGEMENT PROBLEM *(Operations, Design, Construction, Maintenance, Plant Mgmt., etc.)*

(21) TYPE OF ACCIDENT *(Event)*

(22) WHAT WAS USED, DONE, CONTACTED *(Source)*

(23A) HUMAN ERROR *(First selection)*

(23B) HUMAN ERROR *(Second selection)*

(24A) CONDITION DEFECT *(First selection)*

(24B) CONDITION DEFECT *(Second selection)*

(25) REVIEW OF THE MGMT PROBLEM CITED IN SECTIONS D AND E

Sig. ________ REVIEWING MGMT. OFFICIAL

SECTIONS F, G, H, I, and J to be completed by responsible mgmt. analysis identifying problems in the system to be resolved that will reduce accident loss. Keep remarks brief. Use coded information. Leave no blanks. In each section, consultation with supervisor and appropriate mgmt. official is desired.

SECTION F. PERSONNEL PROBLEM *(Consultation with Supervisor and Personnel Official is Desired)*

(26) SUPERVISORY CONTROL AND TRAINING

(27) FITNESS-FOR-DUTY EVALUATION

(28) OPINION OF THE MGMT. PROBLEM RELATED TO PERSONNEL SERVICES

Sig. ________ REVIEWING PERSONNEL OFFICIAL

SECTION G. PROPERTY/EQUIPMENT/ENVIRONMENTAL PROBLEM *(Consultation with Engineering and Property Official Desired)*

(29) MAINTENANCE AND ENVIRONMENTAL CONTROL

(30) FITNESS-FOR-USE EVALUATION

(31) OPINION OF THE MGMT. PROBLEM RELATED TO PROPERTY SERVICES

To be repaired: Yes or No Est. Cost $________

To be replaced: Circle: Yes or No Tentative date: ________

Sig. ________ REVIEWING ENGINEERING OR PROPERTY OFFICIAL

SECTION H. FINANCE PROBLEM *(Consultation with Administrative and Finance Official Desired)*

(32) AMOUNT OF PROPERTY LOSS To Govt. Prop. $________ To "Other" Prop. $________

(33) AMOUNT OF TORT CLAIM AWARD To Government $________ To "Other" Party $________

(34) OPINION OF MGMT. PROBLEM RELATED TO FINANCIAL SERVICES

Sig. ________ REVIEWING FINANCE OFFICIAL

SECTION I. LEGAL PROBLEM *(Consultation with Tort Claim or Legal Official Desired)*

(35) OPINION OF THE PUBLIC SAFETY PROBLEM

(36) POSSIBILITY OF RECOVERY FROM A 3RD PARTY? *(Check One)* __YES __NO *If "No," explain why in (37)*

(37) OPINION OF THE CAUSE, NOT RELATED TO A GOVERNMENT EMPLOYEE OR OPERATION

Sig. ________ REVIEWING TORT CLAIM OR LEGAL OFFICIAL

SECTION J. CORRECTIVE ACTION *(Taken to Make Less Probable the Recurrence of this Accident)*

(38) LOCAL CORRECTIVE ACTION TAKEN OR PLANNED: When: Now ______ Fiscal Year ______

Sig. ________ MGMT. OFFICIAL TAKING ACTION ________ TITLE

(39) RECOMMENDED BUREAU OR DEPARTMENT ACTION TO ASSIST IN SOLVING IDENTIFIED PROBLEMS: *

** A Bureau response is expected if request for action is made here.*

SIGNATURE OF REVIEWING SAFETY OFFICER	DATE	SIGNATURE OF REVIEWING AUTHORITY	DATE	Initial of Bureau Safety Officer	DATE

FIGURE 16-7 Source document from the Department of the Interior *(continued)*

accident report into the company's code. This person must interpret what is written in the report and somehow make it fit an available description that has a code number. This is the beginning of an almost infinite number of possibilities for interesting errors. The person who is doing the actual coding holds the most important key to controlling error. If this person is familiar with the company in which the report originated, the chances of error are lessened, especially if he or she has a technical background. In most cases, however, the coding is done by an inexperienced person in the claims department, usually with no quality control check. Thus at this point it often becomes impossible to ensure accuracy. Some carriers have attempted to remedy this situation by having field safety engineers do the coding, but this has seldom worked well. One company that prides itself on coding integrity because it utilizes field personnel to code found in a recent year that over 40 percent of all claims that were to be coded came in uncoded from the field and had to be coded by a clerk at the home office, using her best judgment. And no attempt was made to determine the accuracy of the 60 percent of the claims that were coded in the field by checking them back with the customer.

If coding is done at the company generating the source document (first report of injury), there is some chance for input accuracy. If coding is done by the insurance company, there is almost no chance. Thus if you are embarking on a computer analysis system and are working with your insurance carrier, retain the coding element yourself; otherwise, you are sure to end up with well-analyzed but unusable information.

GETTING MEANINGFUL INFORMATION

So far we have touched on a number of serious problems we face in dealing with the computer. We have seen that most systems are based on our national standard for analyzing work injuries, Z16.2, which leads us to an analysis of circumstances surrounding accidents, rather than to an analysis of causes. We have also discussed numerous problems in input accuracy. Thus if you are embarking on a computer system, I urge that you deal with these two problems first. If there is an answer to the input-accuracy problem, it lies in (1) getting a source document that will provide answers that are detailed enough to enable you to code the information and (2) having the coding under some kind of tight quality control.

THE U.S. DEPARTMENT OF THE INTERIOR SYSTEM

Figure 16-7 shows an input document that is different from the traditional ones, such as that pictured in Figure 16-1. This document, from the United States Department of the Interior, was devised and implemented by Bill Pope. It has some rather special advantages over traditional input documents. First, it is based on the belief that accidents are symptoms of management problems and operational errors, and it asks the supervisor to look for and record these things, rather than merely the circumstances surrounding accidents. Second, coding is done by the supervisor rather than by someone far removed from what actually happened.

The key to this approach is the analysis made on the back of the form in terms of the identification of the management problem; the personnel prob-

OPINION OF THE MANAGEMENT PROBLEM RELATED TO PERSONNEL SERVICES

Definition: Identifies a specific personnel service that may be related to a cause of accident.

Many human errors are products of how management plans for, receives, develops, and utilizes employees. Work errors can occur because such services need revision to upgrade their utility to the line supervisor.

If there is any reason to connect the personnel services listed below to a cause of accident under investigation, it should be checked off to provide a bank of information that will guide the personnel function.

Code	*Personnel Service*	*Comment*
00	No cause factor related to a personnel service	
01	Task being performed unrelated to that for which he was hired.	Regardless of "other duties" assigned
02	Questionable physical ability for work done	See item (27), performance
03	Recruitment problem	Unfamiliar with locale assigned
04	Labor-management dispute involved	Involved or behind accident
05	Compensation too low for demands of more skill	Imbalance of local wage scale producing accidents.
06	Employee relations problem	Underlying cause of accident
07	Lack of or insufficient training	See item (27)
08	Injury compensation trouble	Problem with handling claims
09	MV licensing problem	SF 46 not issued, not in possession
10	Physical handicap	Placement problem, policy
11	Problem related to classification or qualification	Hazardous work identification
12	Employee or student conduct involved	Disciplinary action arising out of accident
13	Health services needed locally, or insufficient for needs.	First aid facilities needed, enough to support program.
14	Disability retirement involved	Should be considered for this employee
15	Physical examination problem	Employee's phys. cond. to be checked out
16	Dual employment	Cause related to 2nd job, overwork, etc.
17	Check National Driver Register*	Countrywide history of poor driving *
18	Check SMIS driver accident records*	Interior history of poor driving*
19	Local staffing problem	Need more/better qual. employees
20	Hazard duty pay may be considered*	Research to determine the need in SMIS*
21	Communications problem	Between levels of management or between supervisor and employees
22		
23		

*If these codes are entered - be sure to repeat request in Item (39). The Safety Management Information System (SMIS) will be asked for a special printout. It will be sent, through channels, to the originating field station.

FIGURE 16-8 U.S. Department of the Interior's code

OPINION OF THE MANAGEMENT PROBLEM RELATED TO PROPERTY EQUIPMENT/ ENVIRONMENTAL SERVICES

Definition: Identifies a specific property service related to a cause of accident.

Many condition defects are products of how things are purchased, stored, maintained and used. Defects of building structures, machines, materials, tools and the like will produce accidents. Their prompt removal or repair improves the supervisor's chances to operate free from error.

If there is any reason to connect the property services listed below to the causes of the accident under investigation, please check it off in the report.

Code	*Property service*	*Comment*
00	No cause factor related to any property service can be found.	A property official may be most qualified for this decision.
01	Design, specifications, operation changes needed	Unsafe, guards missing, etc.
02	Storage, handling and issue	Bin, bulk, coal and fuel, etc.
03	Building operat., maintenance, and grounds	Substandard bldg. mgmt.
04	Equipment operation, tools, etc.	Repair and maintenance problems
05	Substandard protection program	Fire, personal clothing, mechanical safeguards, etc.
06	Motor Pool Management (GSA) problem	Vehicle condition when loaned
07	Motor vehicle management (Interior)	Official use, defense of suits
08	Purchasing/specifications/safety	Compliance with safety codes
09	Negotiated contracts	Safety clause—"hold safe"
10	Policy, regulations not enforced	
11	Periodic test, inspection needed	Hoists, boilers, elevators, etc.
12	Property disposal problem	Disposal problems—housekeeping
13	Acquisition of excess property problem	Method of acquisition, safety of
14	Space management problem	
15	Board of survey action	Liability for damage problem
16	A non Federal maintenance problem	Icy conditions, holes in sidewalk, etc. *not* under Federal maintenance system. Correction due elsewhere.
17	Check SMIS—may need lift device*	
18	Check SMIS—how many other cases*	
19		See item (37) cause unrelated to Govt. employee/operation.
20		
21	Inadequate fire protection facilities	Lack of water, sprinklers, fire escapes, fire dept. alarm, etc.
22	Inadequate fire prevention activities	Fire inspections, fire drills, fire plan, etc.
23	Need moving equipment	Dolly, cart, etc. to reduce handling
24	A contractor's safety problem	Problem not under control of bureau

SUPERVISORY CONTROL AND TRAINING

Definition: Identifies a problem related to management of people and the training they receive before and after the task assignment.

This item provides the training function with a bank of information about problems related to supervisory control and training.

Analysis of each accident will provide evidence as to the adequacy or effectiveness of past training and will help management in identifying what additional training is needed.

Code	*Supervisory Control and Training*	*Comment*
00	Training and supervisory control is in no way connected with the error or condition defect in this accident situation.	Check this code with care as in most instances, there can be an element of training to be considered.
01	Need a better trained person but unavailable	Understaffed, employment problem, training needs not being met.
02	Emergency task, no time to train	
03	Work pressure allows no time to give adequate training.	Seasonal difficulties, train before sending to field.
04	Close supervision impossible due to circumstances	Great distance between employees, too many people for one supervisor.
05	Skill proficiency needs upgrading, retesting on retraining.	One-time training not enough for the work—provide periodically.
06	Trained employee did not follow what he had been taught.	Local needs growing, not being provided
07	A solo assignment where additional help was needed	Review policy—in view of hazard, two may be a better decision.
08	Work unauthorized	Disciplinary action
09	Written instructions needed	Not a simple matter of supervision
10	Services of prof. safety officer needed	

***If a specific training category can be related to the accident problem in any way, it would help support training needs by checking:*

11	Orientation to-the-job for temporary and regular hires	
12	Management training for supervisors	
13	Technical, scientific, professional training	
14	Skill training for office, clerical people	
15	Skill training for particular trade or craft involved	
16	Safety management for supervisors	
17	Communication between supervisor and employees	
18	Defensive driving	

FIGURE 16-8 U.S. Department of the Interior's code (*continued*)

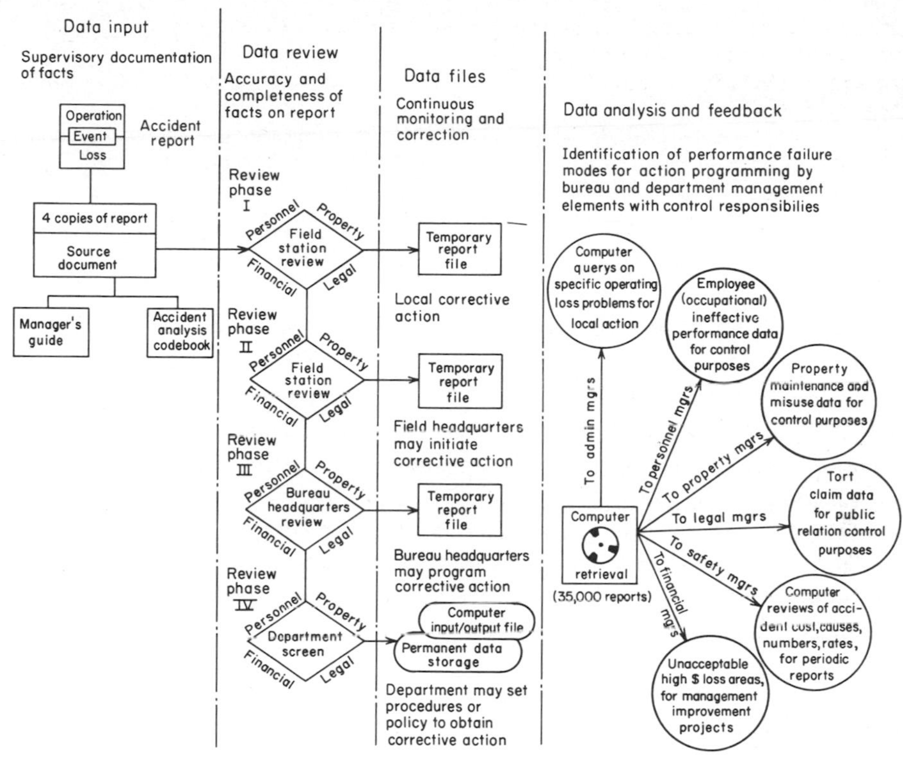

FIGURE 16-9 U.S. Department of the Interior's safety management information system

lem; the problem of property, equipment, and the environment; etc. The codes used for these sections are shown in Figure 16-8.

A flow diagram of the Department of the Interior's safety management information system is shown in Figure 16-9. This is only one of such systems; it is presented in this chapter as an example of a system which is not based entirely on Z16.2 and which attempts to deal with the input-accuracy problem.

Computers have a very real place in our future safety management efforts. However, we shall have to answer some philosophical questions (what do we want to keep in the way of information?) and some system questions (how do we get accurate information?). If we can do this, the computer will become a valuable safety tool.

REFERENCE

Pope, William C.: "In Case of Accident, Call the Computer," in *Selected Readings in Safety,* Academy Press, Macon, Ga., 1973.

17

PROFILING

In Chapters 7 and 8 we discussed an approach to measurement based on an audit of a company or a location for the purpose of determining what is being done that will result in a better safety record. This auditing approach is not really new. It has been widely used in nonsafety situations in many industries with multiple locations. We have also used it in safety. Other countries have used a similar approach, sometimes calling it "profiling."

WHAT PROFILING IS

In profiling, a standard of corporate safety performance in a number of categories considered to be important is developed; then companies are compared with that standard, the end product being a profile showing how a company compares with the standard in a number of categories.

THE CANADIAN APPROACH

In his book entitled *The Industrial Environment,* Jack Fletcher, a Canadian safety professional, presents a method of rating or grading an organization in the areas of injury prevention, damage control, and total loss control. In connection with injury prevention, Fletcher provides a rating outline for such items as loss-control policy, guarding, inspecting, design and purchasing, audio-visual aids, committees and rules, training, investigation, records and analysis, costs, medical examinations, and personal protective equipment. In the area of total loss control, the book presents a way of profiling fire prevention and control, security, health and hygiene, pollution, and product integrity. Fletcher provides a system of rates from 0 to 5, or from unsatisfactory to excellent, and gives the criteria for each rate.

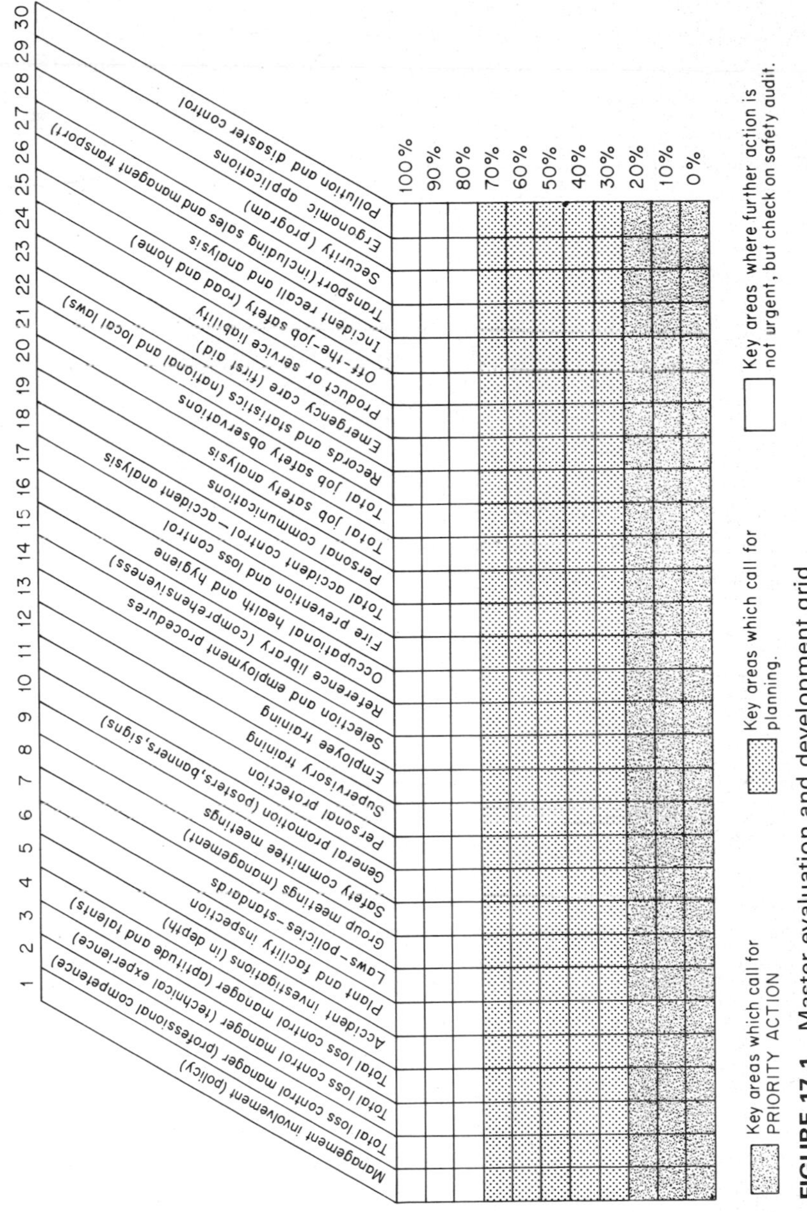

FIGURE 17-1 Master evaluation and development grid

THE BRITISH APPROACH

James Tye, director general of the British Safety Council, offers a similar approach in his book, *Management Introduction to Total Loss Control.* Here he explains his profiling system, in which the profiler rates an organization in 30 different areas and rates the performance on a percentage scale (see his Master Evaluation and Development Grid in Figure 17-1). Each of the 30 areas is then also broken down into key elements. For example, the key area of management involvement has an evaluation and development grid as shown in Figure 17-2.

THE SOUTH AFRICAN APPROACH

Profiling has become the way of life in the Union of South Africa. A profiling approach is utilized by the National Occupational Safety Association (the

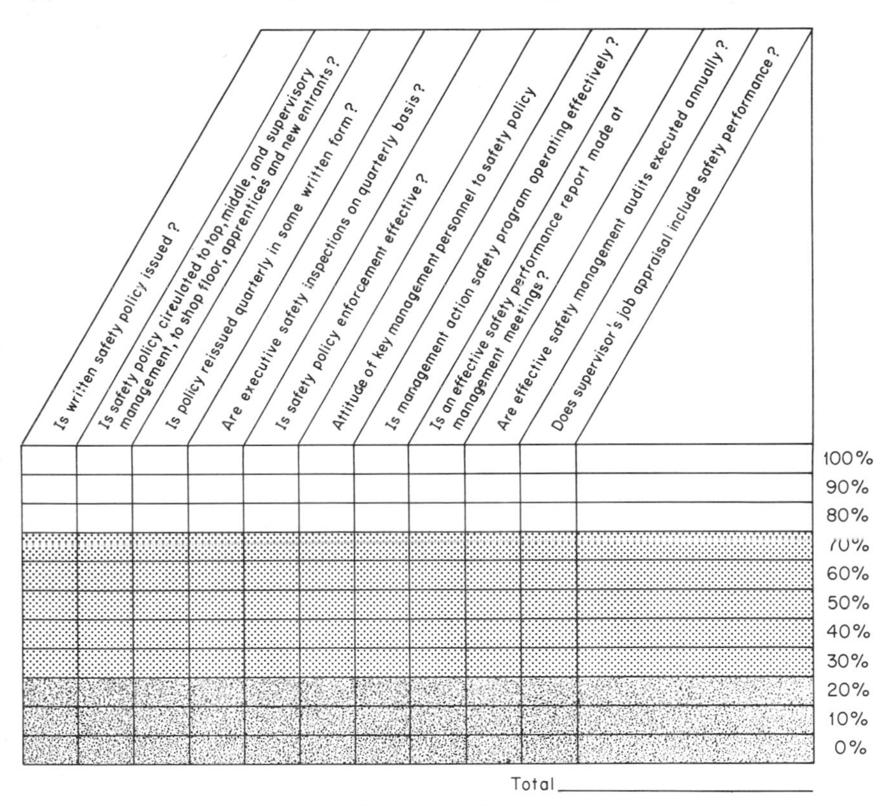

Divide the total percentages by number of questions (10) to obtain average result to carry over to master evaluation and development chart.

FIGURE 17-2 Evaluation and development grid: key area—management involvement

counterpart of our National Safety Council) for grading every industry. The rating report form is shown in Figure 17-3. Bunny Matthysen, general manager of NOSA, has described the system as follows:

Management by Objectives the NOSA Way

The industrialist in South Africa has had much of the preparatory work done for him when it comes to Management by Objectives in the field of Accident Prevention.

Using the best expertise available, NOSA's technical staff, many years ago, arrived at the major objectives required to institute a successful safety program. In broad terms this covered: Industrial housekeeping; electrical, mechanical and personal safeguarding; fire prevention and control; accident recording and investigation; and safety organization. Each major objective was then broken down into subsidiary objectives giving a total of 50 items, which management should strive to obtain.

Some of the objectives were considered more important than others. So in order to differentiate them it was decided to quantify the items. The more important ones were given heavier mark allocations. For example, machine guarding carries 150 marks, whereas clean premises has 40 marks allocated. The total allocation for the 50 items is 2,000 marks.

Under the guidance of the NOSA technical staff, the optimum requirements for each factory can be determined, using the criterion: HOW MUCH MORE could management reasonably be expected to do within the specific plant; taking COGNIZANCE of the materials, the methods, the men and the money available to the plant.

When the safety state of a plant is to be audited by a NOSA technical staff member, the plant starts with a clean sheet and its full quota of 2,000 marks. When deviations from the set objective are observed they are discussed with management and a note is made thereof. Once all 50 items have been investigated in depth the final mark allocation is made by the NOSA man.

From this unique system has evolved NOSA's Star Grading Scheme, whereby recognition for attaining the set objectives is given. To obtain a Five Star Grading the firm must have a mark allocation of over 90 percent. But attaining the objectives must also result in a reduction in the injury frequency rate. Therefore another criterion is added to the requirements of a Five Star Grading—the injury frequency rate must not be greater than five disabling injuries per 1 million worker-hours exposure. At the other end of the scale, in order to attain a One Star Grading, the mark allocation must be between 40–50 percent and the injury frequency rate must not exceed 25.

The NOSA report form no. 4.13.1 depicted on the following page is sent to management after each safety audit where X marks the spot for major problem areas.

Using the NOSA MBO system management can achieve fantastic results in advertising their objectives of conserving our greatest asset—skilled manpower.

Although popular in other countries, profiling is not widely used in the United States. An exception is the approach developed by Frank Bird and used by the Insurance Company of North America for a period of time in working

CONFIDENTIAL

NATIONAL OCCUPATIONAL SAFETY ASSOCIATION

SAFETY EFFORT SURVEY/RATING

Form 4.13.1

SURVEY/RATING OF MESSRS.

NOTE: Items marked "X" require management's attention and should be read in conjunction with the accompanying report. Please refer to the booklet "Management by Objectives" for advice on effective management practice in accident prevention.

Item		Max.	Actual	Action
1.00	PREMISES AND HOUSEKEEPING			
1.10	PREMISES			
1.11	Buildings and floors – clean and in good state of repair	40		
1.12	Good lighting (natural and artificial)	20		
1.13	Ventilation	30		
1.20	HOUSEKEEPING			
1.21	Aisles and storage demarcated	30		
1.22	Good stacking and storage practices	50		
1.23	Factory and yard – clear of superfluous material	60		
1.24	Scrap and refuse bins – removal and disposal	30		
1.25	Colour coding – machines, pipe-lines – other	40		
	SECTION RATING	300	%	
2.00	ELECTRICAL, MECHANICAL AND PERSONAL SAFEGUARDING			
2.10	MECHANICAL EQUIPMENT			
2.11	Machine guarding	150		
2.12	Lock-out system and usage	40		
2.13	Labelling of shut-off valves, switches, isolators	30		
2.14	Ladders, stairs, walkways, platforms	40		
2.15	Lifting gear and records	40		
2.16	Compressed gases; pressure vessels and records	30		
2.20	ELECTRICAL EQUIPMENT			
2.21	Portable electrical equipment – monthly check and records	40		
2.22	Earth leakage relays – permanent and portable	30		
2.23	General electrical installation	50		
2.30	HAND TOOLS – All types: condition, storage and use, e.g. hammers, chisels	50		
2.40	PROTECTIVE EQUIPMENT (Issued: use)			
2.41	Head protectors	20		
2.42	Eye protectors	20		
2.43	Foot protectors	20		
2.44	Protective clothing, including hand protectors	20		
2.45	Respiratory equipment	20		
2.46	Maintenance	20		
2.50	NOTICES – Electrical, mechanical, protective equipment, etc.	30		
	SECTION RATING	650	%	

EFFORT RATING VALUES

NO. OF STARS	RATING %
5* Excellent	91–100
4* Very good	75–90
3* Good	61–74
2* Average	51–60
1* Fair	40–50

Item		Max.	Actual	Action
3.00	FIRE PROTECTION AND PREVENTION			
3.01	Correct types of extinguishers	40		
3.02	Areas demarcated and clear, extinguishers accessible	20		
3.03	Locations marked	20		
3.04	Maintenance of equipment	30		
3.05	Storage flammable material	30		
3.06	Signs to exits: alarm system	30		
3.07	Fire fighting drill and instructions on fire extinguishers	80		
	SECTION RATING	250	%	
4.00	ACCIDENT RECORDING AND INVESTIGATION			
4.10	RECORDS			
4.11	Adequate accident recording (register and dressing book)	30		
4.12	Internal accident report form signed by supervisory staff	30		
4.13	Adequate accident statistics kept in accessible place and NOSA informed	30		
4.20	INVESTIGATION of accidents and remedial measures taken to prevent recurrence	60		
	SECTION RATING	150	%	
5.00	SAFETY ORGANISATION			
5.10	SAFETY PERSONNEL			
5.11	One person made responsible for safety co-ordination by management, in writing, i.e. safety officer, permanent/part-time	30		
5.12	Appointment and acceptance of appointment in terms of Factory Regulation C.7.2(a) and (b) or Mines and Works Regulation 2.9.2	30		
5.13	European Safety Committee	80		
5.14	Non-European Safety Committee or any other similar system	40		
5.15	First aider and equipment	20		
5.16	First aid training	30		
5.20	SAFETY PROPAGANDA			
5.21	Poster programme, bulletins, newsletters, use of safety films and internal safety competitions, etc.	130		
5.22	Notice board indicating injury experience	20		
5.23	Suggestion scheme	20		
5.30	INDUCTION TRAINING AND JOB INSTRUCTION, continuous training e.g. poster appreciation, lectures, rule book. NOSA safety training courses	50		
5.40	PLANT INSPECTION system of reporting to management on safety conditions	50		
5.50	WRITTEN SAFE OPERATING PRACTICES and procedure issued, displayed and explained to the illiterate	50		
5.60	ANY ITEM NOT DETAILED			
5.61	COMPANY POLICY (Also Total Loss Control Programme)	100		
5.62	Bonus points awarded or penalty points deducted			
	SECTION RATING	650	%	
	OVERALL RATING		%	

FIGURE 17-3

with some of its larger insureds. However, not many other uses have been made of profiling in this country.

REFERENCES

Fletcher, John A.: *The Industrial Environment,* National Profile Limited, Willowdale, Ontario, 1972.

Tye, James: *Management Introduction to Total Loss Control,* British Safety Council, London, 1970.

18

SAFETY IN THE ADJUNCT FLEET

The staff safety specialist usually has no defined responsibility in the area of fleet safety. Often no one in the organization has such responsibility. The safety specialist ignores this part of the operation, and fleet management ignores the safety segment of its responsibility. Too often the fleet operation is allowed to go uncontrolled.

Statistics show that in many states (perhaps all states) the motor vehicle is the number one cause of occupational death—the number one industrial killer. Hence, to ignore this problem is to ignore one of the major causes of injury and death. In short, the safety professional cannot afford to ignore the fleet of vehicles that his or her company runs.

This chapter is not intended for the safety professional who works with the true fleet operator, but rather for one whose responsibility extends to a small fleet of vehicles that is adjunct to the primary operations of the company. This might be the contractor's fleet, the manufacturer's pickup and delivery trucks, the fleet of bakery or dairy trucks, or the sales department's private passenger fleet. This chapter gives broad outlines of the accident controls found to be effective with this type of fleet operation.

In dealing with safety in the adjunct fleet, we should keep in mind the basic principles expressed in Chapter 2. Principle 2 states that certain sets of circumstances can be predicted to produce severe injuries. The adjunct fleet offers such a set of circumstances, and statistics show that severity is a distinct possibility, if not a probability. Why, then, do we in safety so often tend to ignore this source of high-potential accidents in our operations?

FAILURE TO DEAL WITH FLEET SAFETY

Lack of Communication

One reason we tend to leave the adjunct fleet alone is that we may be so positioned in the organization that we cannot effectively reach and influence fleet management. Often we in staff safety report to industrial relations or to production management, whereas the fleet reports to an officer who is not closely connected with either industrial relations or production. In some instances the adjunct fleet is part of the sales department. In others it may report to a traffic manager, to the head of material handling, or elsewhere. In terms of the principles expressed in Chapter 4, the safety professional who reports directly to the manager in charge of a particular activity probably will not be able to influence other areas, such as the fleet, easily.

Differences in Fleet Safety Work

Another reason why industrial safety specialists may not extend their influence to the fleet is that they may not feel comfortable in fleet safety work; they may feel that fleet safety is different from plant safety and that the approaches normally used to effect results will not work here. In fact, fleet safety is different from plant safety in two major ways, discussed below.

Lack of Supervision In the plant or on the construction job all employees are under constant supervision. On the road employees are not supervised; they are on their own. This is perhaps the biggest difference between plant safety and fleet safety. Since our supervisory control is so much more limited in fleet safety, we must rely more heavily on some of our other controls—in this case, specifically on deciding who will drive initially. Selection of personnel becomes critical in fleet safety.

The Environment In industrial safety we lean heavily on controlling the environment—on making it as safe as possible. In fleet safety we can control the condition of our own vehicle, but most of the rest of the driving environment is not under our control. The other driver, the condition of the other vehicle, and the road are uncontrollable to us. Here again we lose an important control, and hence we must again lean heavily on some other control that we do have. Much of this lack of environmental control can be balanced by the quality of the driver. Consequently, the selection and training of that driver are of paramount importance.

THE SAFETY PROFESSIONAL'S ROLE IN FLEET SAFETY

Although, as we have said, fleet safety is different in some aspects from industrial safety, the approach of the safety professional remains unchanged. Principle 5 of Chapter 2 reminds us of our role in looking at, and working with, the small adjunct fleet. In many cases the industrial safety specialist should begin by asking whether controls are well established in certain areas. The following are the areas that might be looked at first. Notice the similarity between these areas and those in industrial safety.

1. Management's policy
2. Driver selection
3. Driver training
4. Vehicle maintenance
5. Records

All these are familiar to the industrial safety specialist. Let us look at each one as it applies to the adjunct fleet.

Policy

A statement of management intent is as essential in fleet accident control as it is in plant safety. The policy in fleet accident control may be separate, or it may be an integral part of the total management safety policy, depending on the organization.

In either case the following must be spelled out:

1. Management considers safety on the road important.
2. The corporate safety program will apply to the driver.
3. Employee cooperation is expected.
4. Specific responsibilities for safety have been assigned to the various levels of management.
5. Accountability will be fixed.

Driver Selection

Selection is the single most important control that management has in fleet safety. Proper selection of drivers requires that management determine the abilities and skills of applicants for the driving job. To determine this, it attempts to obtain information on drivers' experience and performance on previous driving jobs, the driver's job knowledge (technical know-how), and attitudes toward safety. Management should also consider drivers' job performance during the probationary period; this is often overlooked.

The specific tools of selection are:

- The application form
- The interview
- The reference check
- The license check

- The physical examination
- Written tests
- Road and yard tests

In the screening process, management's first job is to determine how applicants have lived and worked in the past. This is the best single indicator of how they will live and work in the future. Of the above list of selection tools, those concerning past performance are the most important indicators.

One of the best tools is the application form, since it covers the essential facts:

1. Driving experience—local, long haul, etc.
2. Job permanence record. Does the applicant stay with a job?
3. Responsibility, maturity, and stability.
4. Past safety performance.

The personal interview provides the face-to-face encounter with the applicant. Here the interviewer can appraise the person's knowledge, attitude, character, maturity, etc. The purpose of the interview is to gather facts. In the interview, these ought to be discussed:

1. Driving experience
2. Knowledge and education
3. Knowledge of the vehicle
4. Experience with the vehicle, maintenance, etc.
5. Record—arrests, violations, etc.

After the application and the interview, some reference checks should be made. Checking prior work references establishes the validity of the information obtained and can be done simply and inexpensively over the telephone. The following should be determined:

1. Was the applicant employed by the company as stated?
2. What type of work was he or she engaged in?
3. What was the applicant's absentee record?
4. What was his or her wage record?
5. What were the applicant's reasons for leaving?
6. Would the company rehire this person?

In addition to making the reference check, management should ensure that the applicant has a valid driver's license. Also, a routine check with the applicable state agency for past violations is well worthwhile.

Every applicant should be given a physical examination to determine whether he or she meets the physical requirements for driving. The driving job, perhaps more than most other jobs in industry, requires this physical check. It seems inconceivable that any manager would put an expensive vehicle, and the possibility of a million-dollar lawsuit, into the hands of a person who cannot see or is subject to blackouts. And yet this happens daily in industry, in construction, and in all kinds of business. It happens even in companies with sophisticated safety programs, where those programs do not seem to apply to the salesperson in the company car.

The application, interview, and physical examination are the central part of any selection system. Management would do well to ensure that the potential employee comes through these well before adding some of the "frills."

Various kinds of tests in addition to the above are often valuable. These tests can take many forms and generally include those questions which the company thinks its drivers should be able to answer. Standardized tests are also available. For example, the yard test is used to determine a driver's skill in handling the equipment without going into traffic. Some exercises often included are the parallel park and the alley dock. This test does not predict whether a person will be a good driver—only whether he or she can maneuver the equipment. The road test (in traffic) should be made over a predetermined route which approximates the kind of driving that the applicant will be required to do if hired.

Driver Training

After a driver is finally put on the payroll, he or she must be oriented and then trained. Most of the principles discussed in the section on training as a tool of motivation apply to driver training also. The training given should, as much as possible, aim at defined needs and should accomplish stated training objectives. Objectives are essential to quality training. Basically, the content of a training program is what management wants the workers to know.

Orientation might include:

1. Company policies and practices
2. State, county, and local traffic laws
3. Defensive driving
4. Customer or public relations
5. Concepts of safety

The ongoing training of drivers might include the following:

1. Vehicle operation.
2. Vehicle condition. How to check the vehicle daily and what to report.
3. Use of company forms.
4. Emergency procedures. Roadside warning devices, use of fire extinguishers, proper use of accident report forms, witness cards, conduct at the scene of an accident, etc.
5. Demonstrations. Use of the brake detonator, skill-exercising equipment, and other training devices.
6. Psychophysical testing devices. The validity of these lies in their training value, not in their use as measurement devices. They can be used to train drivers on reaction time, night vision, stopping distance, etc.
7. Films and visual aids.

Vehicle Maintenance

Fleets usually evolve their own systems of preventive maintenance, which are devised to fit their special needs. Maintenance of leased fleets varies according to the contract between lessee and lessor. A fixed-cost or full-maintenance arrangement usually includes maintenance, repair, and insurance. A financed lease is one in which the leasing company supplies the vehicles and the lessor conducts the maintenance. Regardless of the setup, a preventive maintenance system must be used. It is the role of the staff specialist in safety to ensure that whatever the arrangement for maintenance, the program will be effective.

Generally some kind of maintenance record system is necessary to ensure effectiveness. Records should be designed to:

1. Show when maintenance is needed
2. Provide a schedule of what needs to be done
3. Record what has been done
4. Give costs

Usually, the fleet maintenance record system uses the following forms:

1. Driver's condition report—a checklist of things to be checked daily by the driver and an order for necessary corrections (to be sent to the shop)
2. Maintenance scheduling form—a means of ensuring that the shop schedules all vehicles periodically for routine service

3. Service report—a record showing detailed inspection by mechanics when in for routine service and also repairs needed and when made
4. Vehicle history card or folder—a complete history of each vehicle

Probably the best sources for comprehensive record-keeping systems and forms for preventive maintenance are those of the truck manufacturers and oil companies.

Records

Two other types of records are necessary in fleet operations:

1. Driver records
2. Accident records

Driver Records Every fleet needs to keep some records on each driver. The minimum is a file which contains information on:

1. Hiring
   a. Application form completed for screening
   b. Character references
   c. Results of work reference checks
   d. Record of violations and accidents
   e. Results of interviews
   f. Results of tests given
   g. Physical examination report
   h. Information on previous driver training
2. Job performance
   a. Supervisor's reports
   b. Any commendations
   c. Status in company's safe-driving award program
   d. Performance in maintenance functions
   e. Record of training received
   f. Vehicles driven or assigned
   g. Losses of cargo, money, etc.

h. Any road observation reports

i. Any violations of warning notices

j. Accidents

k. Property damage reports

l. Complaints

m. Appraisals

This folder thus contains a brief account of all information pertaining to a driver's performance on a job. A summary record card can be used to coordinate all this information for ready access.

Accident Records Accident records start with the driver submitting an accident report. In the case of serious accidents, investigations are usually made by the company and the insurance carrier. Management may wish to collect, analyze, and summarize these accident reports to determine trends. Often the information and/or reports are evaluated for (1) type of accident, (2) immediate causes, and (3) driver chargeability. Causes are generally classified in terms of driver, roadway, and vehicle. Of primary importance is proper analysis of driver-related causes. Usually, analyses list improper driver actions as:

Failure to yield right-of-way

Following too close

Failure to signal intentions

Speed

Failure to obey traffic signals or signs

Improper passing

Improper turn

Improper backing

Wrong traffic lane

Other

Rate Formulas In fleet safety, as in plant safety, accident records are analyzed to determine how the company's record compares with its own past records and with the records of other companies. A number of different rate formulas can be used.

1. Annual vehicle-accident frequency per million miles: Multiply the annual number of accidents for a given year by 1 million miles and divide by the actual mileage covered in that year.

$$\text{Accident frequency} = \frac{\text{annual no. of accidents} \times 1 \text{ million miles}}{\text{annual mileage driven}}$$

2. Annual vehicle-accident loss rate: Divide gross revenue for one year by the total dollar losses of vehicle accidents for that year.

$$\text{Vehicle-accident loss rate} = \frac{\text{annual gross revenue}}{\text{annual dollar losses from vehicle accidents}}$$

3. Annual accident rate per driver: Divide the total number of accidents for one year by the total number of drivers.

$$\text{Accident rate per driver} = \frac{\text{total number of accidents}}{\text{number of drivers}}$$

4. Annual employee-injury loss rate: Divide the annual gross revenue for one year by employee-injury dollar losses for that year.

$$\text{Employee-injury loss rate} = \frac{\text{annual gross revenue}}{\text{annual dollar losses from employee injury}}$$

REFERENCE

National Safety Council: *Motor Fleet Safety Manual,* Chicago, 1966.

19

PRODUCT SAFETY

In product safety, as in fleet safety, often no one in the company, including the safety professional, has responsibility for controlling potential losses. This is so even though almost everyone in the organization has an interest in product safety: design, quality control, manufacturing, sales, advertising, safety, field service, and legal. This problem has become so acute that something must be done about it.

This chapter is intended to present only a broad outline of the product safety problem. Before we begin discussing product loss-control principles, however, we should look at and define product liability. A knowledge of product liability and its current trends will help us to see more clearly the significance of product loss controls.

PRODUCT LIABILITY

Product liability concerns the legal responsibility of a manufacturer or seller of a product to compensate a consumer who has been harmed by the product. Under present court rulings, product liability falls into one of two basic categories: negligence and breach of warranty.

Negligence

A manufacturer of a product may be liable for harm caused by the product because of a failure to exercise the care of a reasonable person of ordinary prudence to see that the product would do no harm to the buyer: That is negligence.

A manufacturer's affirmative duty to exercise due care includes the making of reasonable tests and inspections to discover latent (hidden or concealed) hazards. Failure to detect what would have been noticed if reasonable inspections and tests had been performed can constitute a basis for liability should a loss occur. That the manufacturer conforms to standard practices of the industry is not necessarily a valid excuse. The jury determines whether such practices are consistent with due care.

In addition, the manufacturer is required to exercise care in planning and designing a product so that it is reasonably safe not only for the purpose for which it was intended but often also for possible uses to which it might be put that were unintended by the manufacturer. The manufacturer is also obligated to provide adequate warning of possible and latent dangers that may be present when the product is put to its proper and intended use.

In the past, only those who bought the product directly from the manufacturer could recover damages if product failure caused them to sustain a loss. Now anyone who suffers a loss from a product which is defective as a result of negligence by the manufacturer can recover.

Breach of Warranty

Failure of the product to perform as represented by the manufacturer constitutes breach of warranty. There are two kinds of warranty: (1) expressed or published, and (2) implied or unwritten. Until recently, courts held that there exists an implied warranty that all food products sold for human consumption are wholesome and fit to eat. The courts ruled that not only the direct purchaser but also the ultimate consumer can recover damages resulting from impure food. Now this concept has been extended to the point where courts have held that manufacturers are liable to anyone—not just direct purchasers—for breach of implied warranty involving any kind of product.

Express warranty is a written statement made by the manufacturer to the purchaser, guaranteeing that the product will or will not perform in certain ways and meet certain standards. Recently, some courts have ruled that advertising and sales literature amount to warranties guaranteeing the product to the ultimate consumer and that any remote consumer can recover damages for breach of these warranties.

It is apparent that the laws dealing with product liability are changing rapidly and that the changes have all tended to increase the liability of the manufacturer. Because of these changes and because of greater public knowledge concerning lawsuits, both the number of product liability cases and the average size in cost to the manufacturer are increasing rapidly. Jury Verdict Research, Incorporated, has reported that during a recent three-month period there were more of such cases in the courts than in the two preceding years combined. The average penalty in product liability cases is more than twice as high as it is in personal general liability cases. Clearly, manufacturers must stop and evaluate what they are doing in the product loss-control area. They must determine the adequacy of their programs and formulate a plan of action for the future.

LOSS CONTROL

In this chapter we shall look briefly at the following areas and suggest how the safety professional might attack problems in each:

1. Management's policy
2. Design and engineering
3. Purchasing
4. Manufacturing
5. Sales and advertising
6. Field service

Management's Policy

Here again, management's stated policy is essential for articulating management's desires, assigning responsibility, and fixing accountability for the control function. The policy on product safety may be issued separately, or it may be part of management's general policy statement.

A stated policy illustrates to all personnel the need for continued emphasis on product loss control. It expresses management's interest and concern in this area, and it commits the organization to goals for improvement. It can establish broad guidelines for areas of accountability and responsibility in specific product loss controls. The written policy serves as the foundation for all the activities and efforts that are conducted to reduce losses in this area. Stated policy must be followed up by stated procedure, that is, specific ways of handling product claims. This is mentioned here because in cases of claims against products, there is usually no past routine which ensures the immediate availability of accurate information needed for defense against the claim, as there is in cases of employee injuries.

Management's policy also assigns basic responsibilities, not only to those in the organization who have an essential part in ensuring that losses are controlled, but also to those who have a part in minimizing the likelihood of a successful lawsuit against the company. This means that the policy assigns responsibility to the line for manufacturing control and to the staff for their part in product safety. For instance, purchasing will have defined responsibility for incoming quality; advertising will have responsibility to ensure that copy is not legally damaging; and legal may have defined responsibility to recheck all advertising copy and service manuals.

Management may wish to center responsibility under one staff specialist or loss-control coordinator. This coordinator serves initially to determine policy and procedures for management and then to coordinate all departments under his or her direction.

Design and Engineering

Loss control is limited in this area because the controls are primarily a function of the qualifications and performance of the design staff. Although the safety professional is not competent to make any judgment on such things, certain questions can be posed:

1. Have written job descriptions and standards of performance been established for engineering and design personnel? Are they regularly reviewed?
2. Does the head of the engineering department know whether each employee is practicing the things necessary to produce a good design?
3. Is all work checked by a superior before release to manufacturing?
4. Do all product failures receive a full analysis so that design and manufacturing can make any necessary changes?
5. Do design and engineering personnel receive training? If so, what is the purpose of such training?
6. Are rigid specifications written for each part? Are drawings and specifications tied together so that purchasing, manufacturing, and quality control follow the desires of the original design?
7. Are all changes in materials and manufacturing approved by design?

Purchasing

In the area of purchasing, the staff specialist can do a good job of analyzing present controls and spotting weakness. Some of the things to observe are:

1. What specifications are there for purchasing?
2. Are the purchasing specifications set by engineering? Are they based on product tests and known standards?
3. How are incoming parts or materials tested?
4. Are proper records kept of these tests?
5. Are materials or incoming parts identified so that they can be traced to the supplier?
6. What procedures ensure that design changes are communicated to purchasing?
7. What is the warranty setup with suppliers?

Manufacturing

In this area, of course, most manufacturers have some product quality controls in effect. The product safety specialist need only ensure that such controls are adequate and are properly in effect. The following might be determined:

1. What kinds of inspections are made in manufacturing?
   ___Trial run
   ___First piece
   ___Pilot piece
   ___Working
   ___Key operation
   ___Sampling
   ___Percentage
   ___Preassembly
   ___Functional
   ___Efficiency
   ___Endurance
   ___Destructive
   ___Product
2. What inspection systems are used?
   ___Patrolling
   ___Before
   ___Centralized
   ___After
   ___During
3. What methods of inspection are in use?
   ___Visual
   ___Gauging
   ___Measuring
   ___Testing
4. Are major parts and assemblies identified with parts numbers? Are records kept of major parts and assemblies?
5. Does the finished product have a key identification number? Where is it located? Will the number stay on the product despite rough usage?
6. What records are kept for production?
7. What type of rejects result from production? Do they have any salvage value?
8. What happens to factory rejects? Are they destroyed? Sold as scrap? Sold to the public at reduced prices?
9. What procedure is followed when a change is made during a manufacturing run? Are dealers or field service notified of changes which may affect service of the product?
10. Are the products certified or inspected by Underwriters Laboratories, American Gas Association, or a similar agency?

Sales and Advertising

As we have said, either an expressed or an implied warranty may go with a product. Occasionally, manufacturers also make use of disclaimers, which are written statements that disclaim or limit the liability for a product. Such disclaimers are of little use with consumer products but can be of value with contracts in which the parties are of equal strength.

When a product has inherent hazards, the laws require that the customer be warned of these hazards. Even when laws are not specific on this point, the company may be liable if warning is *not* given.

The product safety specialist in the area of sales and advertising should not attempt to become a legal expert. Rather, his or her role should be that of a liaison person—one who ensures that legal personnel do, in fact, review all labels, literature, manuals, etc. Some things that should be watched for are:

1. Invalid claims concerning product performance
2. Incorrect specifications
3. Statements to the effect that a product is "perfect," "foolproof," or safe for use "under all conditions"

Field Service

It is important that a communications network be established between the manufacturer and the field salespeople, agents, and retail outlets which handle the product. Often statements made by the salespeople can be damaging to a product claim.

Also, when a loss occurs, the field representative of the manufacturer is the person closest to the customer and hence the one most likely to get accurate information. A system must be set up, however, to ensure that this information is obtained. Some questions that a product specialist might ask when investigating controls in this area are:

1. Does the manufacturer clearly state what maintenance and service are required (for example, in service manuals and owner manuals)?
2. Does the dealer have preparation and installation responsibilities?
3. Is the dealer assisted through factory training and bulletin services?
4. Who does installation or assembly?
5. Are installation or assembly instructions clearly written and, if necessary, properly illustrated?
6. Are adequate warnings given on key aspects of installation or assembly?
7. Is the manufacturer the only source of replacement parts?
8. Are parts lists provided? What is the location of parts depots?
9. Are parts available for obsolete products? What is the source of such parts? Is there a need to replace entire assemblies?

10. What procedure is followed when manufacturing changes require parts to be replaced with dissimilar parts?
11. Are major hazards identified? Are there warnings concerning load limits, electrical grounding, overspeeding, and excessive temperature and pressure?
12. Are operating instructions clearly written? Are major controls identified? Are starting directions, etc., included?
13. Does the manufacturer provide operator training programs?
14. Can recall or field repair be accomplished?
15. What procedures are set up for accomplishing recall or field repair?
16. Is there a standard procedure for dealing with customer complaints?

Loss Procedures

A major role of the product specialist in this area is to ensure that procedures are instituted which allow for the best possible handling of product claims. Such procedures are for the following purposes:

1. To get prompt reports from the field of incidents which may lead to potential product liability cases
2. To get complete information regarding the incident
3. To promptly report the incident, with as much information as possible, to the insurance carrier
4. To assist the insurance carrier in the investigation of the incident
5. To inform company management of the incident and of possible ramifications in regard to sales, design, and manufacturing
6. To assist in the examination of the defective product and to preserve evidence
7. To improve customer relations through prompt action on customer complaints

REFERENCES

Getzoff, Byron: "The New Theory of Product Liability," *National Safety News,* February 1969.

———: "Product Liability: Guides for the Corporate Manufacturing Executive," *National Safety News,* February 1967.

Mazel, Joseph: "Zero Defects," *Factory,* July 1965.

Philo, Harry: "Negligent Design," *American Trial Lawyers,* June 1966.

Robb, Dean: "Safety Is Not Just Common Sense," *Journal of the ASSE,* December 1965.

Schroeder, Richard: "An Insurance Engineer Looks at Product Liability," *Journal of the ASSE,* January 1968.

Zinch, W. C.: "The Foreman and Quality," *Supervision,* April 1966.

———: "A Guide to Zero Defects," Office of the Secretary of Defense, 1965.

APPENDIXES

The first five sections of this book are meant to be read and digested. The appendixes, which contain information not readily available in other sources, are intended to be used by the reader as they serve his or her purposes.

The following are included:

Appendix A: How to Analyze Your Company Two outlines are presented that can help the safety professional or safety consultant evaluate a company's safety performance.

Appendix B: Sample Management Safety Policies A collection of policies are presented that illustrate the principles described in Chapter 3. These policies were selected not because they are good (although some surely are) but rather because the application of each resulted in notable success.

Appendix C: A Special-Emphasis Program The NO STRAIN campaign, aimed at the control of back injuries, is presented as an example of a special-emphasis safety program.

Appendix D: Programmed Instruction This appendix describes a proved technique that is not used enough in safety. A short course in noise control, an area of vital concern in safety today, is also included.

Appendix E: The Safety Game A management game is included in this appendix. It is intended as an example of a different training technique—and can be utilized as such—and it can also be used by the reader as a check on his or her ability to analyze the safety problems of another organization, the fictitious Eastabrook Manufacturing Company.

APPENDIX A

HOW TO ANALYZE YOUR COMPANY

This appendix contains two outlines that can be used to obtain information on a company and evaluate its progress and performance in safety. They are:

1. A loss-control analysis guide. This outline, typical of those used by many insurance companies, can be used to spot-check areas that should be examined and analyzed in order to determine the effectiveness of the current safety program. It is intended to trigger the safety engineer's thoughts—to remind the engineer to look closely at the most pertinent areas of management as they apply to safety.
2. A safety program appraisal. This outline was developed for the purpose of appraising each plant in a multiplant operation in a similar manner. In a multiplant operation with a corporate safety department, it could be used to audit plant safety operations. The outline was compiled from several systems used by a number of different companies.

A LOSS-CONTROL ANALYSIS GUIDE

If we are to understand an organization's safety performance, we must first obtain facts about how accidents are presently being controlled. Next we must analyze those facts to determine where weaknesses exist. Then, in the course of our everyday work or in a special report to management, we can offer suggestions for minimizing those weaknesses. Some of the areas in which to obtain facts are:

I. MANAGEMENT ORGANIZATION

A. Does the company have a written policy on safety?

B. Draw an organizational chart and determine the line and staff relationships.

C. To what extent does executive management accept its responsibility for safety?

1. To what extent does it participate in the effort?

2. To what extent does it assist in administering?

D. To what extent does executive management delegate safety responsibility? How is this accepted by:

1. The superintendent or top production people?

2. The foremen or supervisors?

3. The staff safety people?

4. The employees?

E. How is the company organized?

1. Are there staff safety personnel? If so, are their duties clear? Are responsibilities and authorities clear? Where is staff safety located? What can it reach? What influence does it have? To whom does it report?

2. Are there safety committees?

a. What is the makeup of the committees?

b. Are their duties clearly defined?

c. Do they seem to be effective?

3. What type of responsibility is delegated to the employees?

F. Does the company have written operating rules or procedures?

1. Is safety covered in these rules?

a. Is it built into each rule, or are there separate safety rules?

II. ACCOUNTABILITY FOR SAFETY

A. Does management hold line personnel accountable for accident prevention?

B. What techniques are used to fix accountability?

1. Accountability for results:

a. Are accidents charged against departments?

b. Are claim costs charged?

c. Are premiums prorated by losses?

d. Does supervisory appraisal of supervisors include looking at their accident records? Are bonuses influenced by accident records?

2. Accountability for activities:

a. How does management ensure that supervisors conduct toolbox meetings, inspections, accident investigations, regular safety supervision, and coaching?

b. Other?

C. Are any special systems set up?

1. SCRAPE

2. Other

III. SYSTEMS TO IDENTIFY PROBLEMS—HAZARDS

A. Are routine inspections performed?

1. Who is responsible for inspection functions?

2. Who makes inspections?

3. How often are they made?

4. What types of inspections are made?

5. To whom are the results reported?

6. What type of follow-up action is taken?

7. By whom?

B. Are any special inspections made?

1. Boilers, elevators, hoists, overhead cranes, chains and slings, ropes, hooks, electrical insulation and grounding, special machinery such as punch presses, x-ray equipment, emery wheels, ladders, scaffolding and planks, lighting, ventilation, plant trucks and vehicles, materials handling equipment, fire and other catastrophe hazards, noise and toxic controls

C. Are any special systems set up?

1. Job safety analysis

2. Critical incident technique

3. High-potential accident analysis

4. Fault-tree analysis

5. Safety sampling

D. What procedure is followed to ensure the safety of new equipment, materials, processes, or operations?

E. Is safety considered by the purchasing department in its transactions?

F. When corrective action is needed, how is it initiated and followed up?

G. When faced with special or unusual jobs, how does the company ensure safe accomplishment?

1. Is there adequate job and equipment planning?

2. Is safety a part of the overall consideration?

H. What are the normal exposures for which protective equipment is needed?

1. What are the special or unusual exposures for which personal protective equipment is needed?
2. What personal protective equipment is provided?
3. How is personal protective equipment initially fitted?
4. What type of care maintenance program is instituted for personal protective equipment?
5. Who enforces the wearing of such equipment?

IV. SELECTION AND PLACEMENT OF EMPLOYEES

A. Is an application blank filled out by prospective employees?

1. Does it ask the right questions?

B. What type of interview and screening process is the prospective employee subjected to before being hired?

C. How are the prospective employee's references and past history checked?

D. Who actually does the final hiring?

E. Is the physical condition of the employee checked before hiring?

1. If a physical exam is given, how complete is it?
2. How is the information used?

F. Are any skill, knowledge, or psychological tests given?

G. Are job physical requirements specified from job analysis?

1. Are these requirements considered in new hires?
2. Are they considered in job transfers?

V. TRAINING AND SUPERVISION

A. Is there safety indoctrination for new employees?

1. Who conducts it?
2. Of what does it consist?

B. What is the usual procedure followed in training a new employee for a job?

1. Who does the training?
2. How is it done?
3. Are written job instructions based on the job analysis used?
4. Do they include safety?

C. What training is given to an older employee who has transferred to a new job?

D. What methods are used for training the supervisory staff?

1. How are new supervisors trained?

2. Is there continuous training for the entire supervisory force?

3. Who does the training?

4. Is safety a part of it?

E. After employees have completed the training phases of their job, what is their status?

1. What is the quality of the supervision?

2. What use is made of the probation period?

VI. MOTIVATION

A. What ongoing activities are aimed at motivation?

1. Group meetings, literature distribution, contests, film showings, posters, bulletin boards, letters from management, incentives, house organs, accident facts on plant operations, other gimmicks and gadgets, and activities in off-the-job safety

B. What special-emphasis campaigns have been used?

VII. ACCIDENT RECORDS AND ANALYSIS

A. What injury records are kept? By whom?

B. Are standard methods of frequency and severity recording used?

C. Who sees and uses the records?

D. What type of analysis is applied to the records?

1. Daily analysis

2. Weekly analysis

3. Monthly analysis

4. Annual analysis

5. Analysis by department

6. Cost analysis

7. Other

E. What is the accident investigation procedure?

1. What circumstances and conditions determine which accidents will be investigated?

2. Who does the investigating?

3. When is it done?

4. What type of reports are submitted?

5. To whom do they go?
6. What follow-up action is taken?
7. By whom?

F. Are any special techniques used?

1. Estimated costs
2. Safe-T-Scores
3. Statistical control charts

VIII. MEDICAL PROGRAM

A. What first-aid facilities, equipment, supplies, and personnel are available to all shifts?

B. What are the qualifications of the people responsible for the first-aid program?

C. Is there medical direction of the first-aid program?

D. What is the procedure followed in obtaining first-aid assistance?

E. What emergency first-aid training and facilities are provided when normal first-aid people are not available?

F. Are there any catastrophe or disaster plans?

G. What facilities are available for transportation of the injured to a hospital?

H. Is a directory of qualified physicians, hospitals, ambulances, available?

I. Does the company have any special preventive medicine program?

J. Does the company engage in any activities in the health education field?

A SAFETY PROGRAM APPRAISAL FOR MULTIPLANT OPERATIONS

Appraisal Factors

I. GOALS

A. What safety objectives were set for this period?

B. What progress was made toward achieving those objectives?

II. GROWTH OF PERSONNEL

A. How are we improving the safety knowledge of our personnel?

	Subjects	Time	Effectiveness
Line supervisor			
Intermediate manager			
Plant engineer			
Product engineer			
Industrial engineer			
Plant manager			

III. INSPECTIONS

A. Are inspections effective?

B. General plant inspections made:

Item	When inspected	By whom	Items corrected		Comment
			Last period	This period	

C. Below are some items that might be checked in making general safety inspections:

General layout
Flow of material
Traffic flow
Aisles
Machine controls
Fumes
Illumination
Temperature
Floor loads
Pits and excavations
Areas under construction
Flammable liquids
Noise
Radiation
Fire protection
Guarding
Rest rooms
Washrooms
Access ladders
Stairs
Elevated platforms
Protective equipment
Electrical equipment
Others

D. Preventive maintenance inspections:

Item	Does standard procedure exist?	When inspected	By whom	Effectiveness

E. The following are pieces of typical equipment for which preventive maintenance inspections will probably be necessary:

Elevators	Lift trucks
Hitching equipment	Other trucks
Ladders	Boilers
Scaffolds	Ovens and furnaces
Hoists and cranes	Power presses
Chains and slings	Portable grinders
Crane runways	Stand grinders
Exhaust systems	Power conveyors

IV. PROCEDURES

A. Are regular SOPS effective in these areas?

	Does standard method exist?	Effectiveness
Equipment lockout		
Labeling of containers and pipes		
Welding or burning permits		
Glove issuance		
Approval of equipment overloads		
Entering enclosed spaces, tanks		
Others		

V. EXPERIENCE

A. On-the-job

	Last period	This period	Significance*
Number of industrial first-aid injuries			
Number of disabling injuries			
Number of PPDs and TPDs			
Number of fatalities			
Time lost because of first-aid injuries (estimate in days)			
Days lost or charged because of disabling injuries			
Disabling injury frequency rate			
Disabling injury severity rate			
Combined frequency-severity indicator			
Estimated cost of first-aid cases			
Medical and compensation costs paid out			
Medical and compensation costs incurred			

Cost of property damage and public liability damages for motor vehicle accidents			
Cost of other public liability damages			
Cost of fire losses			
Total industrial injury and accident costs			
Insurance loss ratio			

*The Safe-T-Score could be used here to determine whether the changes noted are statistically significant.

B. Off-the-job

	Last period	This period	Significance
Number of off-the-job injuries (first-aid cases)			
Number of injuries reported to health and accident insurance carrier			
Number of disabling nonindustrial injuries			
Frequency rate for off-the-job injuries			
Time lost from work because of off-the-job first-aid cases (estimated in days)			
Time lost from work because of disabling off-the-job injuries			
Cost of treating nonindustrial first-aid cases (estimate)			
Cost of off-the-job injuries reported to health and accident insurance carrier			
Total cost of first-aid and insurance for off-the-job injuries			

C. Injury summary

	Last period	This period	Significance
Total disabling injuries—industrial and off-the-job			
Total time lost—industrial and off-the-job			
Total industrial and off-the-job injury and accident costs			
Industrial and off-the-job cost factor (per 1,000 worker-hours)			

VI. INVESTIGATIONS

A. What operational factors may be indicators of accident trends?

	Last period	This period	Significance
Accident causes and corrective action:			
Number of incidents reported			
Number where no effective action was possible			
Where action was taken, give number where:			
Guarding was changed			
Work method was changed			
Equipment or facility was modified			
Employee was reinstructed in method			
Protective equipment was changed			
Employee was cautioned			
Employee personal factor was given added consideration			
What causes and agencies were most predominant?			

VII. EXPENSES

A. What has been the total safety program expense?

	Last period	This period	Significance
Total staff expense			
Total cost of protective equipment			
Total safety expense			
Safety expense factor (per 1,000 worker-hours)			

VIII. OUTSIDE ACTIVITIES

A. Are we participating in community safety affairs?

List projects	No. of people involved	Time allotted	Evaluation of project

IX. PLANT MANAGEMENT

A. Describe relationships that help or hinder the program:

X. SUGGESTIONS

APPENDIX B

SAMPLE MANAGEMENT SAFETY POLICIES

This appendix discusses actual management safety policies which have been notably successful in industry. Each policy is included here for a reason. The policies are for the following:

1. Columbia River Paper Company, a division of Boise Cascade Corporation, Boise, Idaho. Here we have a perfect example of a success story starting with policy. This policy was written by a young man who had just been given the job of mill safety director. He was inexperienced in safety, but he learned fast. The first thing he did after accepting the position was to write a policy and submit it to management for its signature. Management signed the policy and backed it. Under the direction of management and the guidance of the safety director, this mill went, almost overnight from a frequency rate of slightly worse than average to a record of 2½ million worker-hours without a lost-time injury.

2. Richardson Paint Company, Austin, Texas. Too often we think of policy as something elaborate. It does not have to be. The letter in this example from the president of the company actually was safety policy—and very effective policy, as it was written by the president himself and spelled out his desires.

3. An unidentified contractor. The construction industry is different. We all know this. But management is management, regardless of the type of operation. Policy is nothing more than management in action, expressing its desire and intent. This policy worked for one contractor. Policy is essential in construction, as it is in any operation.

example I
COLUMBIA RIVER PAPER COMPANY, BOISE CASCADE CORPORATION

The following memo was sent from the newly appointed safety director of this mill to the mill manager. The policy was approved by the manager, starting this mill on a remarkable safety record—over 2½ million worker-hours until its next injury.

TO: ____________________ *Mill Manager*
FROM: ____________________ *Safety Director*
SUBJECT: PROPOSED SAFETY POLICY

Since the job of safety director is a newly created position in this company and since no written policy is available, I felt that my first duty should be to develop a basic outline of company safety policy, for two reasons:

1. Nearly every company with an effective safety program has a written safety policy.
2. This is the best way I know to acquaint you with my basic plan for our safety program.

When I was assigned the job of developing a safety program for this mill, paper mill safety was brand new to me. So I have used the following sources of information in making up this proposal:

The safety seminar put on by our insurance carrier
The Portland staff of our carrier
The Pacific Coast Association of Pulp & Paper Manufacturers
The Camus mill safety staff of Crown Zellerbach Corporation
The Oregon Stage Industrial Accident Commission
The National Safety Council

From these sources I have been able to study the essential parts of the safety policies of probably a hundred mills. This proposal is not patterned after that of any particular company, but is a combination developed from many different programs along with some original ideas of my own.

I am not too optimistic enough to believe that my thinking will dovetail exactly with yours. Therefore, I am submitting this as a proposal only. I would like you to read it over and make whatever deletions, corrections, or additions you feel are necessary. If, in this way, it can be made into an acceptable statement of safety policy for our mill, then it can be used as a guide for future safety activities of everyone concerned. If you think it is workable for us or not at all suitable, or if you should like to discuss any part of it with me, let me know.

Fire prevention and protection are not mentioned because I have not yet had an opportunity to look into that field; and for now I think it would be better to keep them separate.

SAFETY POLICY

COLUMBIA RIVER PAPER COMPANY

INTRODUCTION

A. In order to clarify the safety activities of Columbia River Paper Company, division of Boise Cascade Corporation, the following is set forth as a basic program to clearly establish its existence. It consists essentially of an outline of the relationships of management and employee responsibilities which are necessary for an effective safety program.

B. We wish it to be known that this program will become a basic part of our management policy and will govern our judgment on matters of operation equally with considerations of quality, quantity, personnel relations, and other phases of our management policy.

C. The program is developed and administered, upon approval, by the safety director. In doing so, the safety director is performing a function of the resident manager. His primary purpose is to assist management by encouraging safety consciousness and employee participation. This is accomplished by the development of promotional material, program planning, motivation incentives, safety meetings, inspections, etc.

PURPOSE

The management of Columbia River Paper Company holds in high regard the safety, welfare, and health of its employees. We believe that "Production is not so urgent that we cannot take time to do our work safely." In recognition of this and in the interest of modern management practice, we will constantly work toward:

A. The maintenance of safe and healthful working conditions

B. Consistent adherence to proper operating practices and procedures designed to prevent injury and illness

C. Conscientious observance of all federal, state, and company safety regulations

RESPONSIBILITY

A. The resident manager has taken the responsibility to develop an effective program of accident prevention.

B. Plant superintendents are responsible for maintaining safe working conditions and practices in the areas under their jurisdiction.

C. Department heads and supervisors are responsible for the prevention of accidents in their departments.

   1. They are directly responsible for maintaining safe working conditions and practices and for the safety of all men under their supervision.

   2. Nowhere is the quality of supervision more apparent than in housekeeping. Good housekeeping is not only essential for safety but also indicative of an efficient department.

   3. Each supervisor is responsible for the proper training of the employees reporting to him. Job hazards and safe procedures should be fully explained to each employee before he begins work.

   4. It is also the supervisor's responsibility to see that required personal protective equipment is used in accordance with safety rules and practices.

5. Supervisors should encourage employee safety suggestions and give them immediate consideration.

6. Supervisors will schedule departmental safety meetings as often as necessary to effect safe practices and work methods.

D. Foremen are responsible for the prevention of accidents in their crews.

1. They will enforce all general and departmental safety rules and regulations.

2. They must see that all accidents are reported and that first aid is rendered in case of injury.

3. They will investigate all accidents and near misses and prepare foreman's reports of accident.

4. Each foreman must have a current first aid card in his possession.

E. The safety director is delegated by the resident manager and has the responsibility to provide advice, guidance, and any such aid as may be needed by supervisors in preventing accidents, including:

1. Safety indoctrination of new employees
2. Protective-equipment emphasis (safety shoes, ear protectors, etc.)
3. Safety meeting planning and assistance
4. Supplying information and educational material for meetings
5. Providing forms for safety meeting minutes, hazard analysis, etc.
6. Accident investigation follow-up
7. Suitable job placement for employees able to return after injury
8. Statistical reporting and study.
9. General publicity—handouts, management reports, memos, letters, notices, etc.
10. Arranging periodic safety inspection of the department
11. Scheduling and conducting special safety meetings for all employees
12. Arranging first aid training and other special instruction

F. The safety director will also:

1. Prepare for approval an annual program designed to encourage a broad safety awareness in the plant.

2. Prepare and keep adequate records of all accidents, and from these prepare such charts as will best show the way to eliminating these accidents.

3. Keep in touch with new developments in the field of accident prevention, personal protective equipment, and first aid equipment and procedures, so that he will be an effective guide to everyone connected with the basic safety program.

4. Coordinate the joint efforts of management and labor, and direct these toward totally safe operations.

5. Coordinate development of safety educational material. Prepare and schedule sessions for safety training of supervisors, foremen, committeemen, and other employees.

6. Promote the public relations aspect of the safety program including preparing for approval releases to news media outside the company.

7. Conduct plant safety committee meetings, providing the material necessary for departmental safety meetings.

G. Employees are responsible for the exercising of maximum care and good judgment in preventing accidents.

1. No job shall be considered efficiently completed unless the worker himself has followed every precaution and safety rule to protect himself and his fellow employees from bodily injury throughout the operation.

2. Employees should report to their foreman and seek first aid for all injuries, however minor these may be.

3. Unsafe conditions, equipment, or practices should be reported as soon as possible.

4. Employees are to read and abide by Columbia River Paper Company safety regulations and all departmental safety rules.

5. Employees will be provided with whatever personal protective equipment is necessary—they are expected to use it.

6. Each employee should consider safety meetings as part of his regular job. Reasons to be excused must be just as important as those for missing any of his regular shift.

Following this basic policy and list of responsibilities, there were, of course, lists of other rules and regulations of the company, as well as descriptions of the operation of several safety committees and of several inspection systems.

This policy was accepted by mill management, signed and published by the mill manager, and put into effect. The results were remarkable. Before the existence of this policy, the mill operated with a frequency and severity rate of slightly worse than national average for paper mills. Beginning with the signing of this policy, the company operated for 2½ million worker-hours without a disabling injury. This policy was the starting point for all the activities that went into making this safety record. The author of this policy later became the corporate safety director of Boise Cascade Company, one of the nation's largest corporations.

example II
RICHARDSON PAINT COMPANY, AUSTIN, TEXAS

The following memo from the president was sent to all supervisors of this company. It constitutes policy.

TO: *All Foremen*
FROM: *S. R. Richardson, President*
SUBJECT: SAFETY LETTER

After looking over the large number of reported accidents in October and November and the year-to-date summary, I think it is high time that we pause and do some serious thinking and planning. If our current accident rate keeps up, it is inevitable that someone is going to get seriously hurt or killed.

All of us, whether we are driving a car, walking on the sidewalk or street, working around home, or working on the job, are exposed to hazards of one kind or another. Whether we stay healthy is entirely within our own control, whether we use the small amount of time it takes to recognize a hazard when we see one, whether we are living by the five safety rules, or whether we are only mouthing them.

We have talked thousands of words and written reams of paper to all of you, but the evidence is at hand to show that the necessity of safety is not soaking through. It is high time to quit yakking by mouth or paper and get down to some solid action.

I am making five suggestions. No, suggestion is the wrong word—and will be changed to orders.

1. Immediately upon receipt of this letter, get your crew together and read it to them. You and your men are each to sign and return the attached form immediately. If any man has not studied the safety book and does not know how it is to be used, he is to be laid off without pay until he learns it. Only when he feels that he knows the book from cover to cover, may he return to work—whether it takes one, two, three, four hours or a full week.

2. Until we get this accident situation under control, all safety meetings will be held on the men's own time. We will not pay for them nor are they to be charged to the customer.

3. Even though 99 percent of all accidents are caused by one man's carelessness, nevertheless, unsafe acts that are the basic cause of all accidents are a matter of crew discipline. Each man has a responsibility toward his fellow worker and toward his crew to watch for unsafe acts on the part of a co-worker and call them to his attention.

   In order that each of you realize your individual responsibility, I am giving you two orders:

   a. Whenever a reportable accident occurs where no time is lost, the entire crew will lose 15 minutes time that day, the time to be used to discuss the reportable accident and how it could have been prevented.

b. Whenever a lost-time accident occurs, the entire crew will lose 1 hour for that day, the time to be used in the same manner as in *a* above.

4. Each week, each crew is to discuss safety suggestions, select the best one, and send that one to me signed by every member of the crew. Forms are attached for that purpose.
5. Heretofore, whenever an accident occurred, the foreman and the men injured were to write me a personal letter. That is now being changed. Whenever an accident occurs, the entire crew will use the 15 minutes for only reportable accidents and the 1 hour for lost-time accidents to discuss it with the foreman and the men injured, and the foreman will write up a full report signed by all men and send the report to me.

I know this is tough, but it is not half as tough as we can get unless these accidents come to a screaming halt. They are *not* necessary, and they *are* controllable.

For the past year, we have left accident control and accident prevention pretty much up to the individual, but it is evident that the individual has not assumed that responsibility; and so management must of necessity step back into the picture—and not with padded gloves.

Your life and money are at stake. Your life is much more important to us than our money, and it certainly ought to be of paramount importance to you and your family. Anything that we at the management level can do or that you at the crew level can do is of maximum importance if it will save the life of one man.

I am willing to listen to squawks on these orders, and I am willing to listen to alternative suggestions, but unless the crews can come up with workable plans to take their place, the orders will stick.

It is in your lap, boys. Either come up with some good, workable safety suggestions, or accept the orders as they are written.

We will not put these orders into effect until we have heard from each foreman. If some really good alternative workable suggestion for accident prevention comes in, we will adopt it in lieu of the orders. We will inform you of our decision well in advance and will, if necessary, be willing to discuss the decision with your own appointed committee. But something must be done—and soon.

This memo was sent out from the office of Mr. S. R. Richardson, then president. Attached to it was the return memo. Each supervisor was required to fill it out and return it:

TO: *Mr. Richardson*
FROM: ____________________ *Foreman*

We have had a meeting and have read your letter.

Each of us has the company safety book, and we know the safety rules.

Each of us has not only read but also studied the entire book, and we know the responsibility of each man toward the safety program.

Answer one of the following:

I. We agree that your orders are needed and will follow them:

Yes No

II. We agree with some and disagree with others checked as follows:

1. Yes No
2. Yes No
3. Yes No
4. Yes No
5. Yes No

We have the following alternative suggestions to take the place of the ones we have checked "No."

Date: (Signed) ________________ *Foreman*
Mechanics:

example III
A CONTRACTOR

This policy represents direction in safety by the management of a construction company. It consists of a broad management safety policy, procedures to be followed by superintendents and supervisors, and a set of employee work rules.

MANAGEMENT SAFETY POLICY

It is the policy of this company that every employee is entitled to work under the safest possible conditions for the construction industry. To this end, every reasonable effort will be made in the interest of accident prevention, fire protection, and health preservation.

The company will endeavor to maintain a safe and healthful workplace. It will provide safe working equipment and necessary personal protection, and, in the case of injury, the best first aid and medical service available.

It is our belief that accidents which injure people, damage machinery, and destroy materials cause needless personal suffering, inconvenience, and expense.

We believe that practically all accidents can be prevented by taking common-sense precautions.

Because of the large number of jobs in progress at one time, the varied nature of the work, and the widespread location of the jobs, we must "formalize" our safety program, utilizing written reports and records, to achieve the maximum use and effectiveness of accident prevention information.

To coordinate the safety program, Mr. __________________, Vice-president, will continue to act as safety director. The overall effectiveness of the safety program is his responsibility. His duties include the review and analysis of accident information, safety meeting reports, etc., and the communication of pertinent information to all the jobs and shops.

The responsibility for safety on each individual job and at each shop remains with the superintendent. His duties include the reviewing of all accident investigation and safety inspection reports for the job or shop. He is also responsible for passing safety information along to all his foremen, and for maintaining an accident log to help in identifying accident trends and problem areas so that additional safety effort can be directed as needed. When necessary, he will advise subcontractors, etc., of physical changes or new safety regulations. The superintendent will also conduct regularly scheduled foremen's meetings at which job or shop work progress, hazards, accidents, and other work and safety items will be discussed. A written record of these meetings will be maintained.

The National Safety Council states that "The workmen's attitude toward safety depends absolutely on the attitude of the foreman." The foreman is responsible not only for the quality and quantity of the work produced by the men under him, but for their safety as well. The foreman has the greatest opportunity to create safe working conditions and safe attitudes.

The foreman's daily safety inspections will be supplemented by periodic checklist typewritten safety inspections, as specified by the superintendent.

All work-related accidents requiring professional treatment, and all "near" accidents and accidents requiring first aid treatment where conditions were such that a more serious injury could have resulted, shall be investigated by the foreman; and a written report, in duplicate, shall be submitted, by the foreman, to the superintendent. The foreman will then make sure that the necessary action to prevent similar-type accidents has been taken and/or that others working under his direction are reinstructed or cautioned as needed.

The foreman is responsible for the holding of toolbox meetings with his workers at which accident and other information concerning the safety of his workers and their work is discussed.

The foreman is also responsible for the employees' performance in adhering to the work rules set forth in the employees' "A Safety Policy."

Foremen must advise __________________, Vice-president, two days in advance of any blasting to be done on the job by anyone.

It is wrong to believe that accidents are unavoidable and will always happen. If all of us do our part, including acting and talking safety at all times, a healthy attitude toward accident prevention and an improved safety record can be achieved.

Reducing accident and related insurance costs will permit us to be more competitive in our industry, thus helping to safeguard our jobs.

(Signed) ____________________ *President*
____________________ *Vice-president*
____________________ *Safety Director*

RESPONSIBILITIES

Superintendents

ACCIDENT INVESTIGATION

1. All accident investigation reports will be reviewed by the superintendent to make sure that the proper action has been taken by the foreman to prevent a recurrence of a similar type of accident.
2. Additional action will be directed by the superintendent, as needed.
3. The superintendent will inform the other foremen under his direction, whose operations are such that similar conditions exist, of the accident and of the necessary actions to be taken to prevent similar accidents.
4. The superintendent, when necessary, will advise subcontractors, etc., of physical changes and new safety regulations.
5. The original accident investigation report will be forwarded to Mr. ____________ ____________, Vice-president, for review. The carbon copy will be retained at the job or shop.

ACCIDENT RECORDS

1. The superintendent is responsible for maintaining a record of all work-related accidents and injuries, using the construction-work monthly accident record form.
2. This form will be made out in duplicate.
3. At monthly intervals the duplicate copies will be forwarded to the main office. The originals will be retained at the job or shop.
4. The superintendent will analyze the reports in order to identify any accident trends or problems so that safety effort can be directed as needed.
5. Accidents requiring only on-the-job first aid should be listed and indicated by use of the letters FA in the left-hand margin.

SAFETY INSPECTION

1. The safety inspector's report forms will be reviewed for completeness and accuracy.
2. At monthly intervals these reports will be forwarded to the main office.

SAFETY MEETINGS

1. The superintendent will schedule monthly foremen's meetings at the shops and at the large jobsites, or where conditions warrant.
2. Job or shopwork progress, hazards, accidents, and other work and safety items should be discussed at these meetings.
3. The superintendent will make sure that a written report of the meeting is maintained. The report is to be forwarded to the office of Mr. ________________, Vice-president.
4. When conditions warrant, additional meetings should be held as needed.
5. The superintendent is expected to maintain frequent contact with his foremen to discuss safety and work progress.

Foremen

ACCIDENT INVESTIGATION

All work-related accidents requiring professional treatment, and all "near" accidents and accidents requiring first aid treatment where conditions were such that a more serious injury could have resulted will be investigated by the foreman.

1. Every accident should be investigated as soon as possible to determine the cause and what action should be taken to prevent a recurrence of similar-type accidents.
2. The accident should be discussed with the injured worker after medical treatment has been given, and with others who saw the accident and/or are familiar with the conditions involved.
3. The supervisor's accident investigation report form will be filled out in duplicate. Sufficient information will be included to allow the job superintendent and others to reconstruct the accident so that they can determine whether additional action is needed.
4. Corrective action to prevent similar-type accidents should be taken as soon as possible. It may be necessary to check back to see that this has been accomplished satisfactorily.
5. Both copies of the supervisor's accident investigation report form will be turned in to the job superintendent for review by him and the main office.

SAFETY INSPECTION

1. Each foreman will make a daily safety inspection; and unsafe conditions, including unsafe worker actions, will be corrected promptly. Corrective action should be taken in all areas where a hazard to the worker, physical property, or the public exists.
2. In addition to the above, a weekly written safety inspection will be made, using the safety inspector's report form. As noted above, corrective action will be taken in all areas where a hazard to the worker, physical property, or the public exists. The completed safety inspector's report will be given to the job superintendent for review.

SAFETY MEETINGS

1. Attendance at safety meetings scheduled by the job superintendent is compulsory.
2. Each foreman will conduct a weekly "toolbox" safety meeting with his workers where accident and other information can be discussed. The well-being of physical property and the safety of the public should also be discussed.

Safety Policy for Employees

All employees:

1. MUST wear hard hats on all jobs where there is danger of being struck by falling or moving objects, and on all roadwork.
2. MUST wear safety glasses—when chipping, grinding, operating a jackhammer, drilling above chest height, or whenever an eye injury hazard exists.
3. MUST wear sturdy shoes which are in good repair. Safety shoes are recommended for all persons working on the ground and on decks.
4. MUST use grounded electrical equipment and hand tools which are in good condition.
5. MUST observe all safety precautions and report unsafe conditions to the foreman.
6. MUST understand the foreman's instructions. If you do not know how to do the job safely, ask your foreman.
7. MUST report all injuries immediately.
8. MUST keep their mind on the job; this is the best way to prevent accidents.
9. TRUCK and EQUIPMENT operators MUST be absolutely certain that the path of movement is clear, particularly when they are backing up. The horn or other audible signal must be sounded while backing up.

THIS POLICY SHALL BE OBSERVED BY ALL EMPLOYEES.

(Signed) ____________________ *President*
____________________ *Vice-president*
____________________ *Superintendent*

APPENDIX C

A SPECIAL-EMPHASIS PROGRAM

In Chapter 13, in the discussion of employee motivation, the influence of special-emphasis programs was mentioned. A few such programs were listed, but few details about them were given.

This appendix presents an example of a special-emphasis program: the NO STRAIN campaign of Industrial Indemnity Company, conceived and written by me and by Ray Campbell, former safety analyst of that company. Following is a complete description of the campaign as outlined in the "Manager's Guide," which is an integral part of the campaign packet.

MANAGER'S GUIDE TO A NO STRAIN CAMPAIGN

THE PROBLEM OF BACK INJURIES

Back strains and lifting injuries are among the most frequent and troublesome on the-job injuries. Every material handling job involves a risk, and every employee is potentially likely to attempt an unsafe lift at any time. Strains may be caused by falls, slipping, reaching too far, twisting, lifting too heavy a load, or even repeatedly lifting an acceptable load for a long time.

Whatever the cause, these injuries hurt backs and groins. Such injuries can cause real misery. They also can create feelings of distrust, suspicion, and hopelessness. The net result may be impaired morale, lessened operating efficiency, and higher costs, including raised insurance rates.

In California alone in one year the costs from back and lifting injuries totaled $50 million. This represented about 50,000 injuries—or an average cost of $1,000 per back injury.

WHAT CAN BE DONE?

In many companies chronic frustration over this seemingly insoluble problem creates wrong attitudes that further aggravate the problem. Something can be done. By implementing a NO STRAIN campaign you can take positive steps to eliminate the problem of back and lifting injuries. But the key to the success of the campaign is to begin with the attitude that something *can* be done.

WHAT DOES THE CAMPAIGN CONSIST OF?

The key to the NO STRAIN campaign is to make your employees realize that pain and possible disability are connected with back injuries. Demonstrate by your own attitude how concerned you and your supervisors are with the problem. Connect this idea with the NO STRAIN symbol in your people's minds, and keep them thinking about it through repetition of the NO STRAIN symbol.

KEY ELEMENTS OF THE CAMPAIGN

1. A NO STRAIN folder for employees detailing material handling and lifting techniques [Figure C-1]
2. A NO STRAIN folder for supervisors with facts on material handling, job analysis, and employee motivation [Figure C-2]
3. A NO STRAIN multiple-image wallet card, with rules on correct lifting [Figure C-3]
4. A NO STRAIN lifting instruction chart for use by supervisors in department training meetings [Figure C-4]
5. Two NO STRAIN posters, each 11 by 16½ inches, in red, blue, and black colors [Figure C-5]
6. Three NO STRAIN stickers, each 2 by 4 inches, in red, blue and black colors, for use at points of exposure [Figure C-6]

FIGURE C-1 NO STRAIN employee folder

FIGURE C-2 NO STRAIN supervisor folder

7. Two NO STRAIN banners, each 8 by 60 inches, in red and blue colors, to be hung in each department during the campaign following the kickoff meeting [Figure C-7]

All these materials carry the NO STRAIN symbol—four basically vertical red lines with a sharp angle jog on a blue background. This abstract symbol conveys no specific lifting instruction, but is intended to serve as a reminder to use care when lifting is involved. Display these materials prominently. See that your people learn what the NO STRAIN symbol means.

STARTING THE CAMPAIGN

Do your part. Read the plan of action in this guide. When you understand how the NO STRAIN program works, draw up a plan of attack that meets your needs. Familiarize your supervisors with the campaign, since they will be responsible for implementing it. Study your work methods and the work layout. Make sure that the right equipment is provided for each job. Improved methods increase profits.

Study your training needs. Help your supervisors to encourage proper lifting techniques among all employees.

ENLIST YOUR SUPERVISORS

Enlist your supervisors in the NO STRAIN campaign. Give them the facts they will need to train their men in safe lifting in their jobs. Tell your supervisors what the

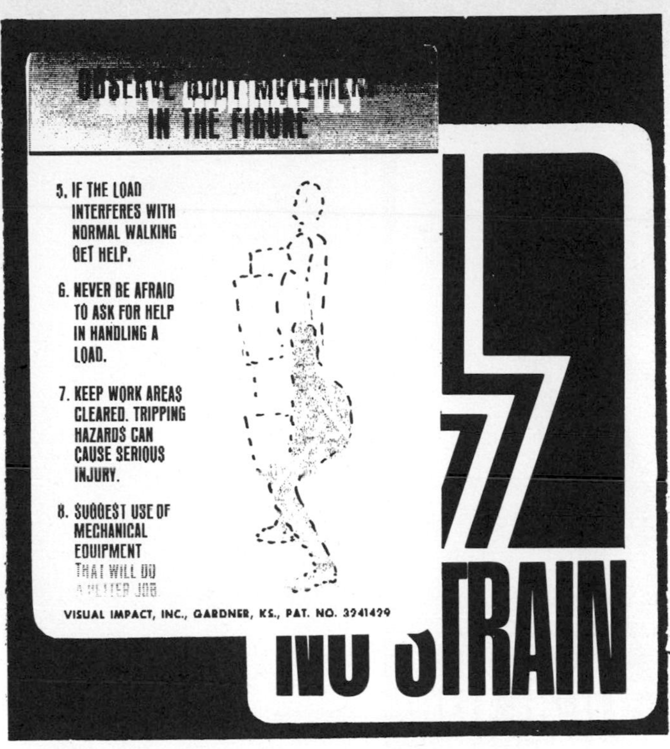

FIGURE C-3 NO STRAIN wallet card

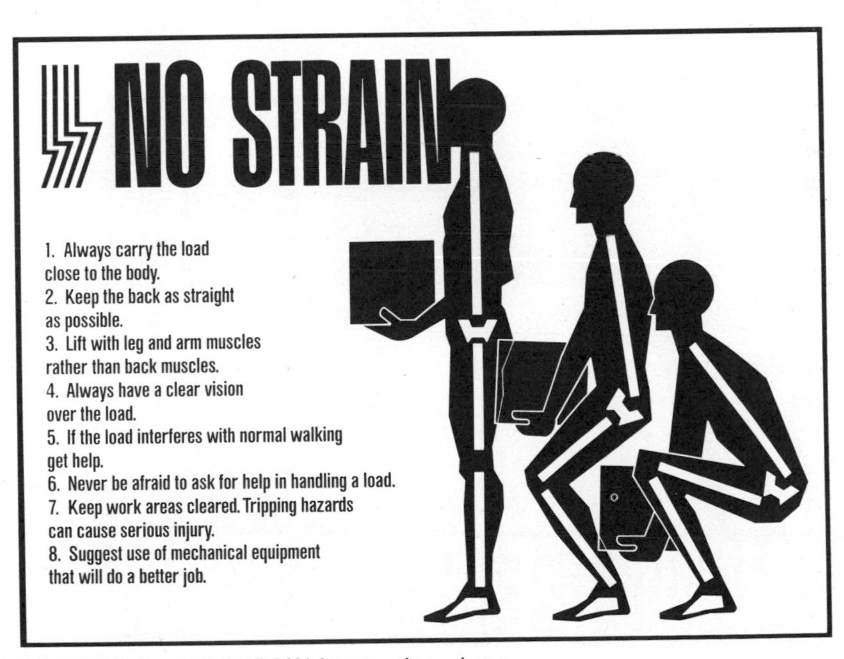

FIGURE C-4 NO STRAIN instruction chart

FIGURE C-5 NO STRAIN posters

FIGURE C-6 NO STRAIN stickers

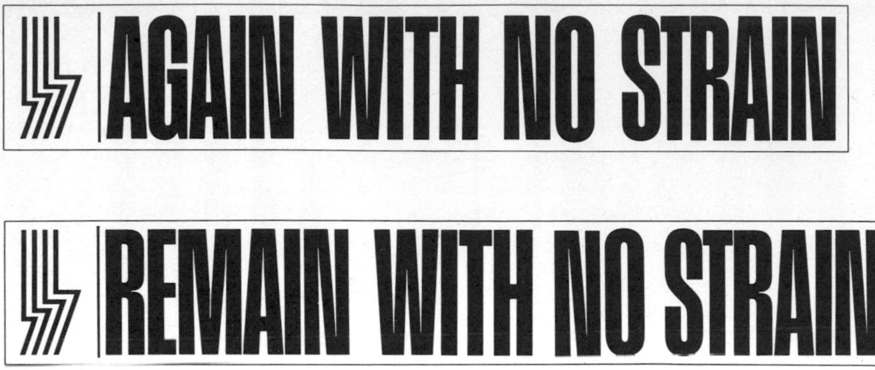

AGAIN WITH NO STRAIN

REMAIN WITH NO STRAIN

FIGURE C-7 NO STRAIN banners

problem is, what you hope to accomplish, how this will help them in their own departments, and what their roles will be. Stress the importance of their giving the campaign their wholehearted support.

Establish an accounting system which will allow you to allocate work injury and material damage costs to each department. Tell the supervisors that you are doing this.

It is essential that the supervisor convey to each employee both his and management's real concern with the problem of back injuries. Each employee must be made to understand that you really care.

Tell your supervisors that you, the manager, will provide them with new employees who are able to work with NO STRAIN, any equipment they need to correct material handling problems, and all the help they need to train and motivate employees in safe lifting. Tell the supervisors that any strains that occur to their employees in the future will be their responsibility, and will affect their own performance evaluations. NO STRAIN can mean higher productivity.

FUTURE HIRES

After the campaign is rolling smoothly and employees have been instructed in the correct methods of lifting, it is important to direct attention to your hiring program. Set up safeguards to ensure that you hire the right man for the job. You will know how much and what kind of strength each job requires if you have analyzed it properly.

Establish a system to find out about prospective employees' problems, if any. Here are the important points to include:

1. A frank discussion of job requirements during the employment interview. Ask the prospective hire whether he has had any back trouble and whether he thinks he can handle the job you are considering him for.
2. A check of previous employers. Find out whether he has had claims for back injuries.

3. A check with the state workmen's compensation board to round out the picture if you are suspicious about his compensation claim history (where this is possible).

By asking the right questions, you should be able to determine which persons are obviously unfit for heavy work. Because some people have less obvious limits on their ability, which only a doctor can discover, you may want to invest in a preemployment or preplacement medical examination. The more completely you describe the job conditions to the doctor, the better he will be able to tell you how capable the person is for the job. In some cases, low-back X-rays may be needed.

When a new employee is hired, start him out right. See that he is properly oriented—told what he can and cannot do, and trained to meet the requirements of the job. This is the place where your supervisors can help.

A SUGGESTED NO STRAIN SCHEDULE

The coordinated NO STRAIN campaign outlined here is intended as a guide, and not as a hard and fast program that must be adhered to without variation. Select a convenient interval between each two of the various phases of the campaign. This plan has been set up on a weekly basis, but you can schedule the individual steps at shorter or longer intervals if it better suits your requirements. Regularity, however, is important; so draw up a schedule and stick to it.

BEFORE THE CAMPAIGN

Order supplies of all forms, folders, posters, stickers, and banners that you will need for the program. Explain the employment procedure, medical examination program, and use of the NO STRAIN interview guide to your employment interviewers. Establish an accident cost-by-department system.

Write your supervisors a letter telling them about the problem of back injuries, describing the need for the NO STRAIN campaign, and enlisting their support. Announce the NO STRAIN kickoff meeting. This kickoff meeting should be scheduled for week 1 just after the supervisors receive the letter.

WEEK 1

Mail the NO STRAIN letter to employees so that it will reach them before the department kickoff meeting. Hold the supervisors' meeting. Issue the materials to supervisors. Tell them to make individual contacts with employees. Instruct the supervisors to put up the posters and banners in their departments. Let each supervisor know that you will be checking with him. Tell him when to give you his first checklist report. Explain that the NO STRAIN campaign will include these activities:

- A NO STRAIN folder will be given to each employee. As the supervisor hands out the folders he must impress each employee with the harsh reality of pain and costly work time lost if the employee injures his back.

- NO STRAIN posters will be displayed on bulletin boards, and banner posters above aisles.
- Each supervisor will be issued enough NO STRAIN point-of-exposure stickers for all his employees—to be given out at the meeting held in week 2. Each employee should be instructed to place the sticker where he thinks it will be most effective as a reminder on safe lifting.
- Each supervisor will receive a NO STRAIN checklist with space for employee names and space for him to list each campaign activity. The supervisor should fill in the date when each activity is completed for each employee and submit the finished form to management.
- Each supervisor will receive an instruction chart showing the correct way to lift. He will hold a meeting with his crew to announce the purpose of the campaign and to describe and demonstrate proper lifting techniques.
- Supervisors should be instructed to make any changes in work method, equipment, or employee placement that are necessary to eliminate a dangerous condition.

WEEK 2

Make sure that your supervisors hold department meetings and that they know how to instruct their crews in proper lifting techniques.

WEEK 3

Either telephone or see each supervisor. Discuss his first NO STRAIN work analysis. Ask him about equipment needed and decide how to get it. Continue to emphasize the company's sincere desire to substitute mechanical equipment for manual handling wherever possible; to take decisive action in the reporting, treatment, and investigation of every back strain; and to hire only people who are able to work with NO STRAIN.

WEEK 4

Review job strength inventories with your supervisors. Remind them to make individual contact with each employee and to issue a second NO STRAIN sticker.

WEEK 5

Begin the first of a series of monthly reviews with each supervisor. Discuss the first month's checklist with him to see how well he has followed the intent of the NO STRAIN campaign.

Hang the second banner (Again with NO STRAIN) and the second poster. Repeat all activities. Give special recognition to supervisors who have done a good job the first month.

TO SUM UP

As a manager you can do something about strains if you:

- Begin with the attitude that something can be done.

- Instill a sense of responsibility in your supervisors for carrying out the program.
- Plan work methods.
- Plan work layout.
- Hire the right people.
- Train your supervisors.
- Get your people to think.

PERMANENT NO STRAIN ACTIVITIES

Supervisors should continue training employees and promoting safe lifting after the campaign. Analysis of work methods, careful job placement, and individual employee

No Strain Checklist

This checklist is to assist you, the supervisor, by reminding you of your role in administering the NO STRAIN campaign. The campaign depends heavily on the supervisor. You must do certain things with your team in order for the campaign to succeed. When these things are accomplished, you should so state on this checklist. Send it to management following the fifth week of the campaign and as often after as management desires.

Department (or craft) ______ **Supervisor** ______

I. Supervisors' kickoff meeting ______ **NO STRAIN banner up** ______

NO STRAIN poster up ______ **(Fill in completion dates)**

Individual contacts (folder to each employee)

Employee		Date	Employee		Date

II. Meeting of employees (card to each) ______ **(Date held)** ______

Follow-up contacts

Employee		Date	Employee		Date

III. Second banner and poster up ______ **(Date)** ______

Work analyses completed **IV. Job strength analyses completed**

Material or part		Date	Job		Date

V. Return this checklist to management.

1E202-SS-NS

FIGURE C-8

No Strain Letter To Supervisors

(Put the letter on your own stationery, preferably your personal memo stock. Make it your own personal message.)

Dear_______________ (Supervisor's name):

The health and morale of our employees, our productive efficiency and our operating costs, including insurance rates, have all suffered from the wrong attitude about back injuries and strains.

We don't accept the argument that nothing can be done. Instead, we're going to start a NO STRAIN campaign and do something about eliminating back and lifting injuries.

The first, and most important, thing is attitude -- yours and mine. Back injuries are real even if you can't see them. The pain is real. The disability is real. The disruption of work in your department is real.

And so are the causes. Among them are (1) jobs that require excessive effort for material handling due to lack of proper hoists and equipment, (2) incorrect lifting methods, and (3) wrong employee attitudes -- chance-taking or showing off. Furthermore, the problem is aggravated by hiring new employees who have a history of "bad backs" -- who are physically disabled.

We're going to do something about these causes in the NO STRAIN campaign. We're going to analyze everything we do to make sure we have substituted machines for people on all heavy lifting jobs. We're going to train our people to use the equipment properly, stop taking chances, and start lifting correctly. And we'll be more careful about who we hire in order to avoid putting a man with a weak back on a job that requires much lifting.

I know we can lick this problem. And I'm counting on you to take full responsibility to see that we do. You control the people who work for you and the material handling methods in your department. I'm going to help you in every way I can to make this campaign a success.

Let's all pull together and work to make back injuries a thing of the past.

FIGURE C-9

contacts should become good habits with them. You should devise your own program for continually following up your supervisors to see that they carry out their responsibilities.

Although the intensive period of the NO STRAIN campaign lasts for only a short time, obviously the concepts of lifting safely are always valid. Refresher meetings should be scheduled at regular intervals in the postcampaign period to ensure that your employees continue to use correct material handling methods. The practice of careful interviewing to determine "back histories" of prospective employees should have become a basic part of your hiring process.

Continue to involve your supervisors in analyzing jobs for accident traps within their departments. Encourage your supervisors to solicit suggestions from their crews on how to correct problems in their work environment. Assign individual workmen to act as safety investigators to bring to light problems that employees may be reluctant to discuss with their supervisors. During the follow-up period every effort should be made to strengthen the safe attitudes developed in the NO STRAIN campaign.

Good luck with your NO STRAIN campaign. Let us work to ensure that back and lifting injuries soon become a thing of the past.

No Strain Letter To Employees

(When you send this letter out, remember that supervisors are employees, too. Be sure that they get a copy.)

Dear Employee:

If you've ever suffered the pain of a back injury or strain, you know how serious these injuries can be. A moment of thoughtlessness can cause injury or even disability -- leading to helplessness, loss of time from work and severe pain.

Let's face it, we've all been guilty of taking chances when it comes to lifting. The problem is not an easy one to solve. Often when a person injures his back, he feels hopeless and misunderstood, and is afraid to admit he's been hurt. This only makes a bad situation worse.

Now, however, we're going to do something about back injuries. With your help we can -- and will -- eliminate the causes by starting a NO STRAIN campaign. We must all recognize how serious strains are, and what they can do to a person, even though only the injured man can feel the pain.

Help us help you by reporting all lifting or material handling problems you know about in your department. If you have a "bad back," don't keep it a secret.

We know that strains can be avoided, but it's up to you at the "moment of truth." You must decide when your back is too tired, when the load is too much, when the reach is too far, or the position too awkward. Don't risk injury -- get your supervisor to help you. You'll earn the respect -- not criticism -- of those you work with. And you'll avoid the pain and discomfort of a serious injury.

In the days ahead you'll hear a lot about the NO STRAIN campaign. With your concerned interest and help, we can all work together to make back injuries a thing of the past.

FIGURE C-10

Job Strength Inventory

(To be completed for each lifting job and sent to doctor who examines new employees.)

Physical Requirements

What part of average work period does employee sit? Give percentage. ____

What part of average work period must employee stand? Give percentage. ____

How much lifting does the job require? Give details. ____

What is the maximum weight the employee must lift? ____ Percentage of total ____

What is the weight the employee must lift most often? ____

How much lifting is:
(Give percentages) At floor level? ____ Waist high? ____ Shoulder high? ____

How much lifting is overhead? Give percentage. ____

How much carrying does the job require? ____

How much turning or twisting is required? ____

How much reaching is required? ____

Special Requirements

Give details of any special requirements of the job such as high work, work in enclosed spaces, crawling, climbing or kneeling.

Recommended Job Improvements

Evaluate the requirements of the job. Is there an easier way to do the job? Consider all possible new layouts and mechanical equipment which could reduce manual material handling operations. List your recommendations:

1E203-5S-NS

FIGURE C-11

Sample forms and letters

Following are samples of the forms, checklists, guides, and other materials that are available as a part of the NO STRAIN campaign, should you wish to make use of them:

1. A NO STRAIN campaign checklist for charting the progress of the program [Figure C-8]
2. A letter for your supervisors telling them what they must do in the campaign [Figure C-9]
3. A letter for employees telling them about the campaign and what will be expected of them to make the program a success [Figure C-10]

4. A job strength inventory to help you analyze the special requirements of each job [Figure C-11]
5. A work-analysis data sheet to help you find and eliminate work-area hazards [Figure C-12]
6. A NO STRAIN interview guide for hiring new employees [Figure C-13]

The NO STRAIN campaign is a fairly complete special-emphasis program, packaged and ready for use by management. Special-emphasis programs have been most effective. Not only will a well-devised and well-run campaign reduce numerically the type of injury being attacked, but it will also measurably reduce all other types of injuries at the same time. This has been true in industry in most cases where these campaigns have been conducted.

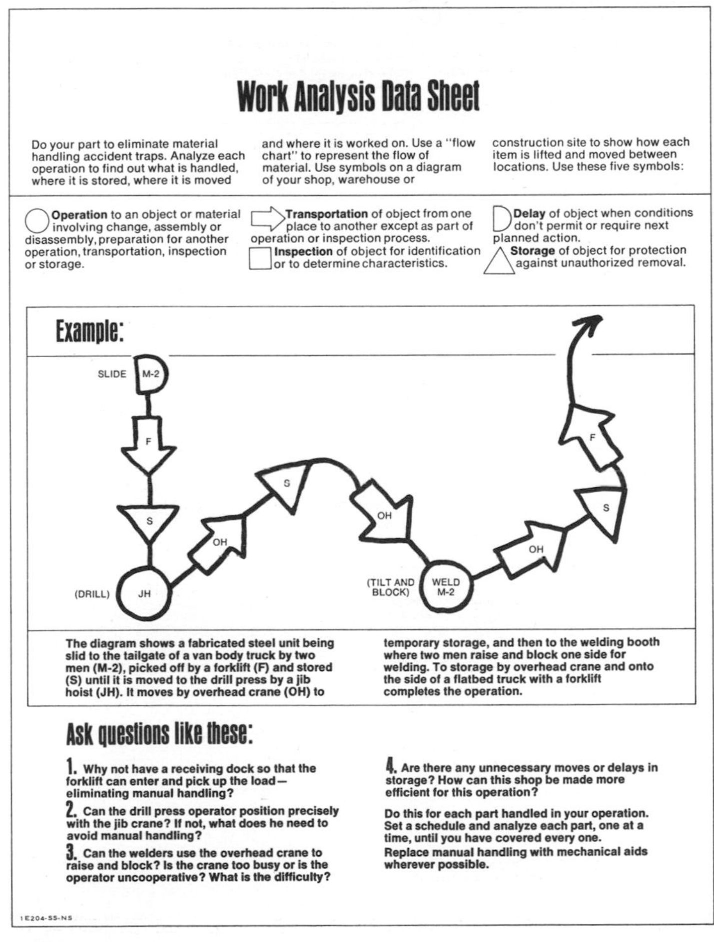

Work Analysis Data Sheet

Do your part to eliminate material handling accident traps. Analyze each operation to find out what is handled, where it is stored, where it is moved and where it is worked on. Use a "flow chart" to represent the flow of material. Use symbols on a diagram of your shop, warehouse or construction site to show how each item is lifted and moved between locations. Use these five symbols:

Operation to an object or material involving change, assembly or disassembly, preparation for another operation, transportation, inspection or storage.

Transportation of object from one place to another except as part of operation or inspection process.

Inspection of object for identification or to determine characteristics.

Delay of object when conditions don't permit or require next planned action.

Storage of object for protection against unauthorized removal.

Example:

The diagram shows a fabricated steel unit being slid to the tailgate of a van body truck by two men (M-2), picked off by a forklift (F) and stored (S) until it is moved to the drill press by a jib hoist (JH). It moves by overhead crane (OH) to temporary storage, and then to the welding booth where two men raise and block one side for welding. To storage by overhead crane and onto the side of a flatbed truck with a forklift completes the operation.

Ask questions like these:

1. Why not have a receiving dock so that the forklift can enter and pick up the load—eliminating manual handling?

2. Can the drill press operator position precisely with the jib crane? If not, what does he need to avoid manual handling?

3. Can the welders use the overhead crane to raise and block? Is the crane too busy or is the operator uncooperative? What is the difficulty?

4. Are there any unnecessary moves or delays in storage? How can this shop be made more efficient for this operation?

Do this for each part handled in your operation. Set a schedule and analyze each part, one at a time, until you have covered every one.
Replace manual handling with mechanical aids wherever possible.

1E204-55-NS

FIGURE C-12

KEY ELEMENTS OF A SPECIAL-EMPHASIS PROGRAM

The following are the key elements of a true special-emphasis program:

1. A defined problem. The program should be aimed at a particular defined problem, usually the one that is causing the most trouble. Examples are falls, motor-vehicle accidents of a certain type, backing accidents, burns, and hand injuries. The problem should be defined on the basis of your company's previous experience.

No Strain Interview Guide

Instructions: Have a frank discussion with the prospective employee. Explain the lifting requirements of the new job from the Job Strength Inventory. Try to get candid answers from the applicant on the lifting requirements of his past jobs. Find out if he has had any back injuries or strains which could be a problem on the new job. Phone the previous employer(s) to verify the lifting requirements and injury experience. Record the replies. Ask the applicant to sign the medical exam approval (bottom of page) before completing the rest of this form.

Last name	First	Middle	Age	Sex	Height	Weight

Health History (Include dates and other pertinent details.)

Have you had any serious falls? ______ Any auto accidents? ______

Back or spinal operations? ______ Back muscle spasms? ______

Chronic backache? ______ Other back problems? ______

Any time on light work or off work during the last two years? ______

Have you received any compensation awards? Give amounts. ______

Work History (Include addresses and phone numbers from employment application. Start with the most recent job and work back at least two years.)

Last employer ______ Maximum weights handled ______

Location ______ How often handled? (Hours daily) ______

Phone ______ How high lifted? (Waist, shoulder) ______

Type of job ______ Date phoned to verify ______

Comments ______

Next-to-last employer ______ Maximum weights handled ______

Location ______ How often handled? (Hours daily) ______

Phone ______ How high lifted? (Waist, shoulder) ______

Type of job ______ Date phoned to verify ______

Comments ______

(Use other side if necessary)

I have no objection to a medical examination or having X-rays taken.

______
Applicant's signature Date

FIGURE C-13

2. A measurable base period. You should have a past record which indicates the problem and gives you something against which to compare results.
3. Measurable results. Records should be kept to make sure you know the results of your efforts.
4. Defined responsibilities. Responsibility for certain activities must be fixed—usually to the line organization.
5. Procedures to fix accountability. Management must know whether the line is carrying out those activities.
6. Planned steps. Decisive steps should be planned before the start of the program.
7. A recall symbol. You should have some symbol that can be seen everywhere and will remind employees over and over again of the program.

APPENDIX D

PROGRAMMED INSTRUCTION

Programmed instruction attempts to teach by breaking the material down into small, "bite-size" portions and giving it to the student one bite at a time. After taking the bite, the student must make a response to demonstrate understanding. After making the response, the student is asked to check to make sure that it was the proper one. The student is then rewarded with the knowledge that he or she responded correctly and is asked to take another bite.

Proponents of programmed instruction have shown that students learn fast and retain what they have learned well. Safety professionals should know about programmed instruction and use it where it fits. It is just beginning to be used in accident prevention—only in recent years have good programs been available.

To illustrate the technique of programmed instruction, the first few sections of a programmed textbook on industrial noise control appear on the following pages.

Instructions

Take a piece of paper and place it across the page just under the statement, "Thus, sound is increased and then reduced. . . ." Now read statement 1 on the right side of the page. In the second paragraph fill in the blank with the word "pressures." Now slide the page down, exposing the answer (1. pressures) on the left and statement 2 on the right. Read and complete statement 2; then slide the paper down again. Continue in this manner through the section.

Section 1
NOISE: ITS CHARACTERISTICS AND MEASUREMENT

---

1. Sound, in nontechnical terms, is variation in air pressures, above and below atmospheric. It is caused by a vibrating body or by a series of pulses of air, steam, or other gas.

Thus, sound is increased and then reduced air __________, which form sound waves.

---

1. pressures

2. Sound waves are set up by any __________ body. They travel through the air at approximately 1,120 feet per second, in all directions.

---

2. vibrating

3. Sound waves are quite similar to the water waves that are produced by dropping a stone into a pond. The main difference is that water waves travel only along the surface, being circular in shape; sound waves spread out in all directions, being __________ in shape.

---

3. spherical

4. Noise may be defined as "unwanted sound." Is music sound? __________. Is it noise? __________.

---

4. Usually
Rock and roll?

5. Two basic elements of noise are: (1) its intensity or loudness: (2) its frequency or pitch.

Intensity is __________. Frequency is __________.

---

5. loudness
pitch

6. If intensity and frequency are within a certain range, they produce the sensation of hearing. This range is known as the range of __________.

---

6. hearing, or
audibility

7. The intensity of noise is measured by a sound-level meter, more commonly called a "noise meter." A sound-level meter measures the __________ of noise.

---

7. intensity, or
loudness

8. A sound-level meter consists of a microphone, a calibrated attenuator, and an indicating meter. The microphone picks up the __________. What does the meter do? __________ __________.

---

8. noise
Visually depicts the
level

9. The unit of sound measurement is the "decibel" (abbreviated dB), which is 1/10 of a bel, named in honor of Alexander Graham Bell. The higher the number of decibels, the louder the sound. A level of zero decibels represents roughly the weakest sound which can be heard by a person with very good hearing.

There are __________ decibels in a bel. 60 decibels is __________ than 50 decibels.

---

9. 10
louder

10. One decibel represents the weakest sound audible. A whisper would read about 10 decibels on the __________. The rustle of leaves is perhaps 20 decibels, or __________ bel. The quiet home or private office is about 30 decibels.

---

10. noise meter, or sound-level meter 1/5	11. A home with children might run 60 decibels (when they are quieter than usual). The average street noise is perhaps 70 decibels, and a noisy office could be 80 decibels. What decibel reading would a sound-level meter read at your home? ________.
11. 115 decibels in mine	12. Above 80 decibels can be considered loud noise. A noisy factory will run over 90 decibels. A boiler factory might run 105 decibels; artillery in action might run close to 120 decibels. Over 120 decibels is the threshold of feeling. Above this ________ level pain could result.
12. decibel	13. Let's take one minute to review. Sound is ________ Noise is ________ Intensity means ________ Frequency means ________
13. variations in air pressure unwanted sound loudness pitch	14. Intensity is measured in ________. A ________ is ________ of a bel.
14. decibels decibel 1/10	15. One decibel is ________. ________ decibels is the threshold of feeling. Most sounds lie somewhere in between. A decibel is actually the measurement of a sound-pressure or intensity level. It is a ratio between the sound-pressure level of the noise being measured and a standard, or reference sound-pressure level.
15. the lowest sound audible 120	16. As with any ratio, this can be expressed by a mathematical formula. A decibel is a logarithmic ratio and is expressed by the formula: $\mathrm{dB} = 20 \log \frac{P_1}{P_2}$ 20 is a constant in the formula, P_1 is the sound-pressure level being measured, and P_2 is the standard, or reference ________ ________.
16. intensity sound-pressure level	17. In actual noise measurement we do not have to worry about this ratio or this formula, since all noise measuring instruments measure directly in decibels. We should, however, understand what the decibel is. A decibel is a ________ between the ________ ________ of the noise being measured and a ________ sound-pressure level.
17. ratio sound-pressure level standard, or reference	18. We said that intensity is ________ and ________ is pitch. The sound-level meter measures only sound pressure or intensity. For a complete description of the sound, measurements involving frequency are also necessary.

---

18. loudness
frequency

19. Frequencies are laid out in octaves in the same manner as a piano keyboard. The term "octave," taken from ________, means the interval between any two sounds having a frequency ratio of 2 to 1.

---

19. music

20. The frequency of 256 cycles per second (or hertz) is approximately that of middle C on the piano, an octave higher is about 512 hertz, the C two octaves above middle C is about ________ hertz, an octave up from that is ________ hertz, and one more octave, which is the top note on the piano keyboard, is ________ hertz.

---

20. 1,024
2,048
4,096

21. The series of frequencies most commonly used for noise-frequency measurements is:

37.5–75 hertz
75–150 hertz
150–300 hertz
300–600 hertz
600–1,200 hertz
1,200–2,400 hertz
2,400–4,800 hertz
4,800–10,000 hertz

Each of these is a ________.

---

21. frequency octave

22. Frequencies are expressed in hertz, abbreviated Hz. 20 hertz would be ________ than 2,000 hertz.

---

22. lower

23. The important frequency ranges are shown below:

Range of audibility						
Low vibration frequency		Most important speech range				Ultrasonics
20	500	1,000	2,000	5,000	10,000	20,000

Frequencies in hertz

From this chart it appears that the most important speech range lies between ________ and ________ hertz.

---

23. 500
2,000

24. 20 to 500 hertz is the low ________ range. 500 to 2,000 hertz is the ________ range. 2,000 to 20,000 hertz is audible, above the speech range. Above 20,000 hertz is inaudible.

---

24. vibration
speech

25. The sound analyzer is used to obtain measurements involving sound frequency.

Intensity is measured by ________.
Frequency is measured by ________.

---

25. sound-level meter
sound analyzer

26. Both instruments, the sound-level meter and the sound analyzer, must be used to measure any noise adequately. Besides the intensity, the ________ must be determined.

These two instruments when used together will give the intensity in each frequency of any noise.

---

26. frequency

27. Most noises are not pure tones. They are composed of several or of many frequencies. Each frequency may have a different intensity.

A pure tone has only ______ frequency. Most noise has several or many ______.

---

27. one
frequencies

28. For instance, the noise from a tumbler might be measured with a sound-level meter and found to be 111 decibels. This reading is called the "overall noise level."

The overall noise level is measured with a ______.

---

28. sound-level meter

29. Now to determine what frequencies the noise is composed of, the ______ must be used.

The sound analyzer might break down the overall noise level of 111 decibels for the tumbler as follows:

The noise level in the 20–75 Hz frequency range—72 dB

75–150 Hz	69 dB
150–300 Hz	74 dB
300–600 Hz	83 dB
600–1,200 Hz	102 dB
1,200–2,400 Hz	106 dB
2,400–4,800 Hz	106 dB
4,800–10,000 Hz	104 dB

---

29. sound analyzer

30. The noise from the tumbler is actually composed of several noises in various frequency ranges. The noise in each frequency range has a measurable intensity.

The intensity is higher in the ______ frequencies than in the ______ frequencies on the tumbler.

The noise then would be considered a high-frequency noise.

---

30. high
low

31. The overall noise level measured with a sound-level meter cannot tell you whether a noise is a high-frequency or a low-frequency noise.

A sound analyzer can do this by measuring the ______ in each ______ range.

---

31. intensity
frequency

32. The intensities of each frequency range add up logarithmically to the overall intensity level.

Thus, the logarithmic sum of the noise levels measured in each frequency range of the tumbler is 111 decibels, the ______ noise level. Figure D-1 shows some typical noise levels in industry.

---

32. overall

Type of machine	*OA*	*37–75*	*75–150*	*150–300*	*300–600*	*600–1,200*	*1,200–2,400*	*2,400–4,800*	*4,800–9,600*
Wood planer, 3/8 inch cut on pine	108	80	77	102	94	100	105	98	93
Cut-off saw	106	83	80	82	88	89	87	104	102
Drum sander, 54 inches	85	84	79	75	73	73	70	69	58
Shaper, on millwork	92	88	87	81	88	87	84	82	74
Cut-off saw, 15 inches	90	88	83	77	75	77	76	79	74
Jointer, idling	99	80	86	98	92	80	76	74	68

Figure D-1 Some typical noise levels

Section 2
NOISE: ITS EFFECT ON HEARING

33. Although there is still much to be learned about the relationship of noise exposure to hearing loss, we have accumulated some information through experimentation and research. For instance, we know that many noise exposures can produce permanent hearing loss. We know that noise-induced hearing loss may be permanent or transient. Transient hearing loss is only temporary; hearing will be restored after the person has been away from the noisy atmosphere for a period of time.

In the case of permanent hearing loss due to noise exposure, hearing will not be restored. There is physical destruction of certain ear structures. Also, the amount of hearing loss produced by a given noise exposure will vary from one individual to another. Loss may result in one person and not in the next, even when the second person is exposed to the same noise. We also know that noise-induced hearing loss first affects our hearing of sounds higher in frequency than those necessary for communicating speech, that is, above the frequency of 2,000 hertz.

Different types of noise have different effects on the hearing mechanism. It is generally believed that noise of high intensity is more injurious to hearing when it is of high frequency than when the frequency is lower. A diesel locomotive horn, for example, produces a sound of low frequency, while the air hose produces a sound of high frequency.

Noise-induced hearing loss depends upon noise levels and exposure time. Any attempt to assess the need for hearing conservation must take account of both.

A person who is exposed to high-intensity noise for a 40-hour week will be more likely to suffer damage than one who is exposed to the same noise for only an hour per day.

The limits for injurious noise levels are still vague and uncertain, and the entire problem of hearing loss due to industrial noise requires additional research and evaluation. The answers to such questions as "How much is too much?" and "How often is too often?" are not yet available. We do know, however, that when certain noise levels are present, it is advisable to initiate a hearing conservation program and a program of noise control.

When the noise is distributed more or less evenly throughout the eight octave bands, or frequency ranges, and when the exposure is regular—many hours each day and five days each week—and if the noise level at 300 to 600 hertz or at 600 to 1,200 hertz is 85 decibels or more, it is time to consider control.

The OSHA safety and health standards state it this way:

> Protection against the effects of noise exposure shall be provided when the sound levels exceed those shown in [Figure D-2] when measured on the A scale of a standard sound-level meter at slow response. When noise levels are determined by octave-band analysis, the equivalent A weighted sound level may be determined as follows:
>
> Octave-band sound-pressure levels may be converted to the equivalent sound levels by plotting them on this graph and noting the sound level corresponding to the point of highest penetration into the sound-level contours. This equivalent sound level, which may differ from the actual sound level of the noise, is used to determine exposure limits from [Figure D-3].

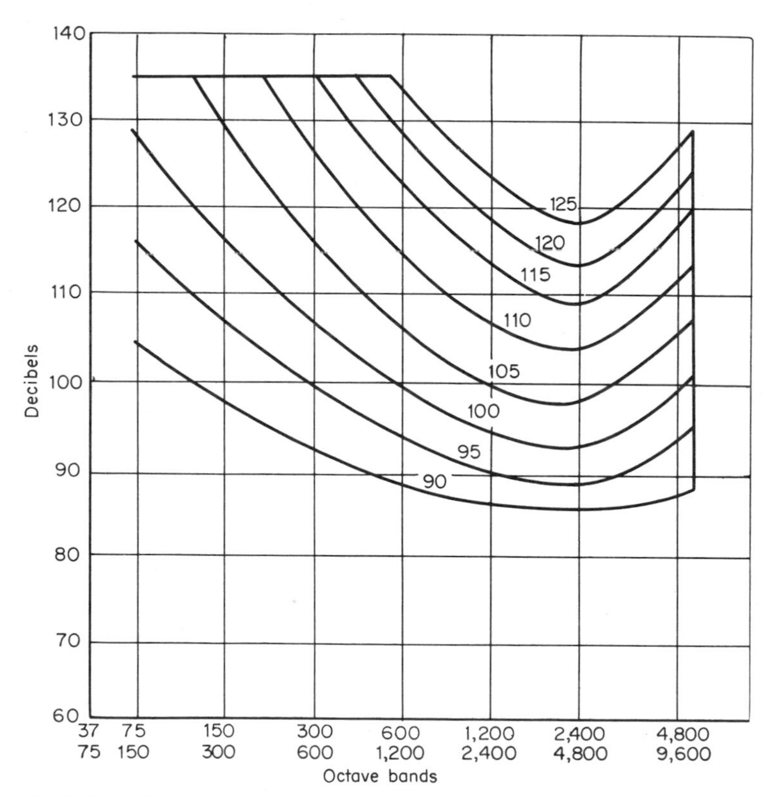

FIGURE D-2 Noise-level contours

Decibels	*Hours*
90	8
92	6
95	4
97	3
100	2
102	1.5
105	1
110	0.5
115	0.25

FIGURE D-3 Permissible exposure times

> When employees are subjected to sounds exceeding those listed in [Figure D-3], feasible administrative or engineering controls shall be utilized. If such controls fail to reduce sound levels to within the levels of the table, personal protective equipment shall be provided and used to reduce the sound levels accordingly.
>
> In all cases where the sound levels exceed the value shown herein, a continuing effective hearing conservation program shall be administered.

To illustrate how this approach works, Figure D-4 plots the tumbling barrel in item 28 on page 260. From the chart in Figure D-4, we can see that while the actual overall noise level of the tumbling barrel is 111 decibels, the equivalent sound level of the noise is also about 111 decibels, and from Figure D-3 we can see that without protection the maximum exposure time is less than 30 minutes per day.

In this case, the actual overall level and the equivalent level are the same. This is not always true. Figure D-5 shows the equivalent chart for a circular cutoff saw. This saw has an overall reading of 106 decibels. However, when plotted, its equivalent level is over 110 decibels, reducing exposure time permitted from one hour a day to less than 30 minutes.

The factors involved in hearing loss are:

A. Loudness (over 85 decibels) and time of exposure

B. The individual exposed and the duration, intensity, and frequency of the noise

C. Distribution, regularity, and individual differences

Turn to page 266 and read the section which corresponds to your answer: A,B, or C.

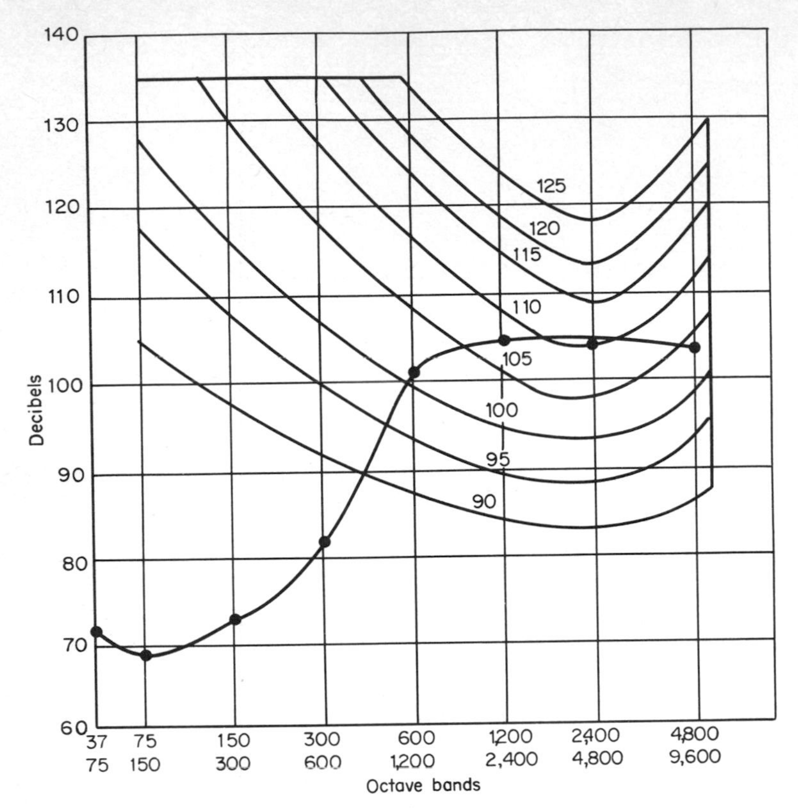

FIGURE D-4 Tumbling barrel noise

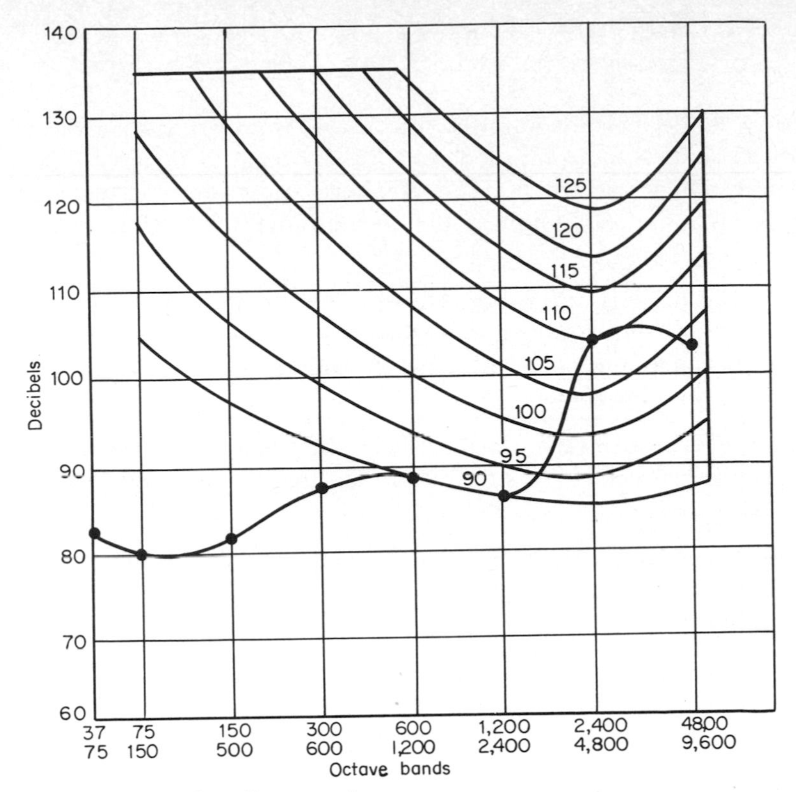

FIGURE D-5 Cutoff saw noise

You answered A: Loudness (over 85 decibels) and time of exposure. It is true that these are factors, but they are not all the factors involved. Let us review:

Loudness (over 85 decibels) or intensity is a factor. However, if the overall noise level of 85 decibels was composed in such a way that the intensity of the 300-to-600-hertz range and the 600-to-1,200-hertz range was considerably below 85 decibels, noise level might not be in the range where controls are needed.

Remember: If the noise is distributed more or less evenly throughout the eight octave bands, if the exposure is regular, and if the noise level at the 300-to-600-hertz range and at the 600-to-1,200-hertz range is 85 decibels, then it is time to consider control.

You said time of exposure was a factor. True, the person exposed for 40 hours per week is more likely to suffer loss than the one exposed for five hours per week.

Now how about individual differences? Two persons will not necessarily react in the same way to noise. One may suffer hearing loss at a given intensity level, and the other may not.

And, finally, what about frequency? Remember the diesel horn and the air hose. The diesel horn makes a low-frequency noise, and the air hose makes a high-frequency noise. Noise of high frequency is more likely to produce loss than noise of low frequency.

You answered B: The individual exposed and the duration, intensity, and frequency of the noise. You are correct. All four are factors.

You answered C: Distribution, regularity, and individual differences. Why did you answer this? Up to now we have not even used the terms "distribution" and "regularity." Maybe a rereading of Section 2 would help.

Move on to Section 3.

Section 3
NOISE: EVALUATING THE EXPOSURE

---

34. The initiation of a hearing conservation program should be considered whenever persons have:

1. Difficulty in communicating by speech while they are in the presence of noise.
2. Head noises or ringing in their ears after working in the presence of noise for several hours.
3. A loss of hearing that has the effect of muffling speech and certain other sounds after several hours of exposure to the noise. (This hearing loss is transient and usually disappears in a few hours.)

The first step to take, before attempting necessary control measures, is to survey a plant in an effort to evaluate the noise hazard. An approximate idea of the noise intensity may be obtained by walking through the plant with another person having normal hearing and trying to carry on a conversation with this person. Where it is difficult or impossible to understand a loud voice at a distance of 1 foot, a noise hazard may be assumed to exist.

If you cannot communicate adequately at ________, no measurement is really needed. You know you have a noise problem.

---

34. 1 foot

35. It should be determined which machines are producing excessive noise, and it should also be determined which noises are necessary and which are unnecessary. Necessary noises are those which are an inevitable result of productive processes, such as a noise produced at a machine's point of operation.

The impact noise of a punch press is an example of ________ noise.

---

35. necessary

36. Unnecessary noises are those which are generally due to faulty maintenance, worn parts, or other causes which can be remedied without involving any fundamental change in operations.

The rattle of a loose guard is ________ noise.

---

36. unnecessary

37. As an important aid in evaluating the noise, reference should be made to available technical data on the average noise levels of the common tools of production in use by all the ordinary types of industry.

Such a reference appears in Figure D-1.

What is the average noise level of a sander? ________________.

What is the average noise level of a planer? ________________.

---

37. About 85 decibels
About 108 decibels

38. It should be determined how many employees are engaged in operations involving excessive, necessary noise and how many others in the adjoining area are exposed to it.

For example, five punch presses may be directly exposing the five operators and indirectly exposing 25 additional employees in the department who themselves are not working on a noisy operation.

To the press operators the press noise is ________ noise.

To the other 25 employees the press noise is ________ noise.

When a thorough survey has thus been made and information such as the preceding has been obtained, proper control measures can be more readily and accurately determined.

38. necessary unnecessary	39. Proper control measures fall into two categories: 1. Medical controls 2. Engineering controls _____________ controls and _____________ controls must work together. Both are essential to a hearing conservation program.
39. Medical engineering	40. To be effective, a hearing conservation program should include: 1. A noise exposure analysis—evaluation of the exposure 2. Provision for control of noise exposure—engineering controls 3. Measurements of hearing _______________ —medical controls
40. acuity	

Section 4
NOISE: THE HEARING MECHANISM

41. The ear is a complex organ which serves as a source of communications and as protection. It also has another function—it governs our sense of balance, which is so necessary in moving about in our environment.

We usually think of the ear as having three parts: the outer ear, the middle ear, and the inner ear. In order to explain the hearing process as simply as possible, we shall discuss each part separately and explain its construction and the work it accomplishes.

The External, or Outer, Ear

Most authorities agree that animals are helped by large external ears. In human beings the outer flaps of the ears do little to aid hearing. The sound enters the ear and goes by way of a passage,' about 1½ inches long, directly into the eardrum, which measures nearly ¼ inch in circumference. The presence of earwax or cotton, for example, in the passageway can prevent sound waves from reaching the eardrum, causing a conduction type of deafness.

The Middle Ear

The middle ear begins at the inside of the eardrum. Attached to it are the three smallest bones of the body, which carry the sound into the inner ear. These small bones, called the "hammer," the "anvil," and the "stirrup," amplify sound as well as protect the ear from loud noises which are apt to be injurious. The stirrup is attached to another membrane, called the "oval window."

There are also two tiny muscles in the middle ear which help to prevent the transmission of too powerful vibrations to the inner ear, at the same time holding the bones closely enough together so that the effect of weak vibrations is not lost.

The lower part of the middle ear chamber has an opening called the "eustachian tube." It has a very important function—it ventilates the ear chamber and equalizes pressure. The air pressure in the middle ear cavity is equalized as the eustachian tube opens during swallowing and yawning.

Have you ever suffered ear discomfort on a plane trip? If so, the air pressure on one side of the eardrum was not equal to that on the other. Adenoids and tonsils likewise can obstruct this air-pressure equalizing mechanism, causing a conductive type of hearing loss. It is this type of conduction deafness which the schools can do much to correct through systematic measurement programs, now being carried out in many states. A great number of individuals have this type of deafness.

The Inner Ear

The inner ear, guarded by a door, the oval window, is filled with a fluid encased in a bony structure. A sensitive membrane of 2½ turns containing the sensitive organ of Corti makes up the most complicated portion of the hearing mechanism. Approximately 30,000 nerve endings are located on this membrane, all connected together into one nerve carrier, the eighth cranial nerve, which carries the sound waves to the brain, where they are interpreted.

A sound wave, carried through the air, reaches the outer ear and travels into

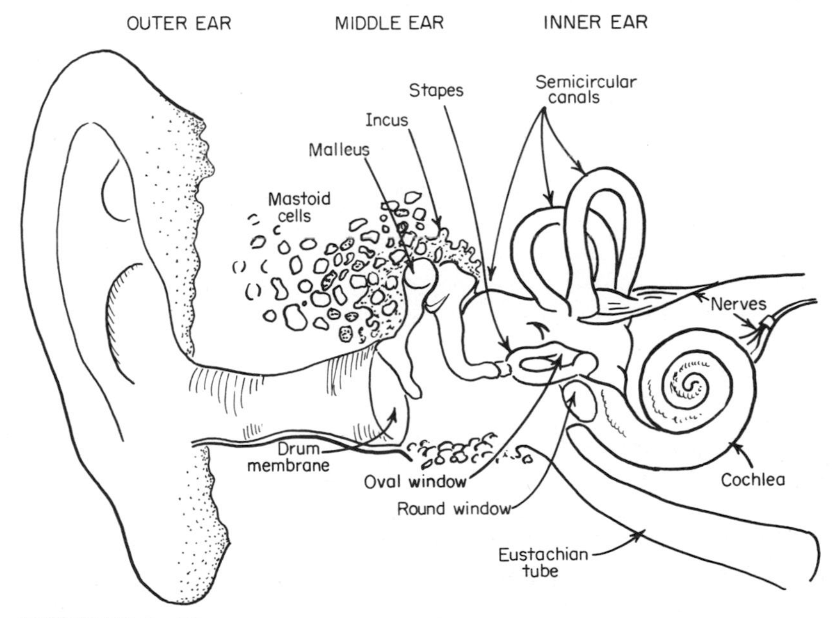

FIGURE D-6 The ear

the auditory canal, where it strikes the eardrum. This moves the small bones, which carry it through the open space in the middle ear until it finally reaches the inner ear by way of the oval window. The vibrations rock the fluid in the inner ear, causing certain nerve fibers to move, depending on the type of sound, and these fibers transmit the sound to the brain via the eighth cranial nerve. Tones of different frequencies stimulate the hair cells on the membrane in different regions; high frequencies stimulate those near the bottom end, and low frequencies stimulate those near the top of the organ of Corti.

Failure of the nerve mechanism to register and transmit sound to the brain is called "perceptive deafness" or "nerve-type deafness."

Receptive losses, or nerve losses, are located in the high-frequency area.

Conduction losses are located in the low-frequency area.

The ear consists of three parts. They are:

A. Canal, lobe, and eardrum

B. Outer ear, middle ear, and inner ear

On page 271 read the section which corresponds to your answer: A or B.

You answered A: Canal, lobe, and eardrum. Perhaps you should read Section 4 again,

You answered B: Outer ear, middle ear, and eardrum. Correct. Now move on to Section 5.

Section 5
NOISE: MEDICAL CONTROLS

	42. Most early noise-induced hearing losses pass unnoticed unless they are detected by suitable hearing tests. A suitable hearing test can be defined as a pure-tone air-conduction threshold test. The term "air conduction" describes the path by which the test sounds reach the ear. The test is generally referred to as an "audiometric examination." The testing machine is an audiometer. An audiometric examination tests ________. It is performed with an ________.
42. hearing acuity audiometer	43. In audiometric exams, the test sounds are generated by earphones and conducted through the air in the ear canal to the eardrum. Recommended test frequencies are 500, 1,000, 2,000, 3,000, 4,000, and 6,000 hertz. The record of results is called a "threshold audiogram." An audiogram records ________ in an audiometric exam.
43. results	44. Audiometric measurements should be given to all new employees before they begin work. An audiogram is shown in Figure D-7. An audiogram is the permanent ________ of the test.

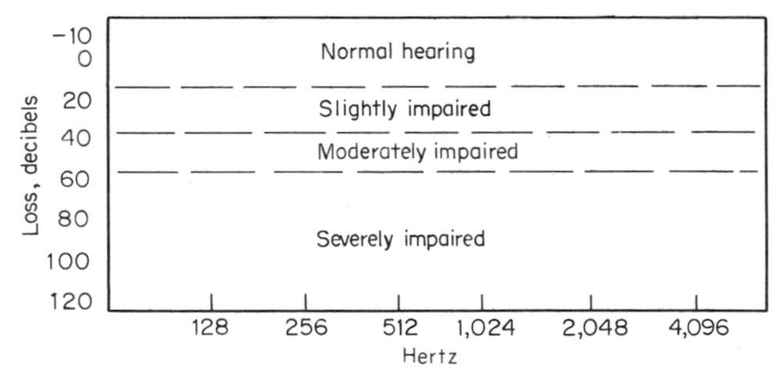

FIGURE D-7 Audiogram showing levels of hearing loss

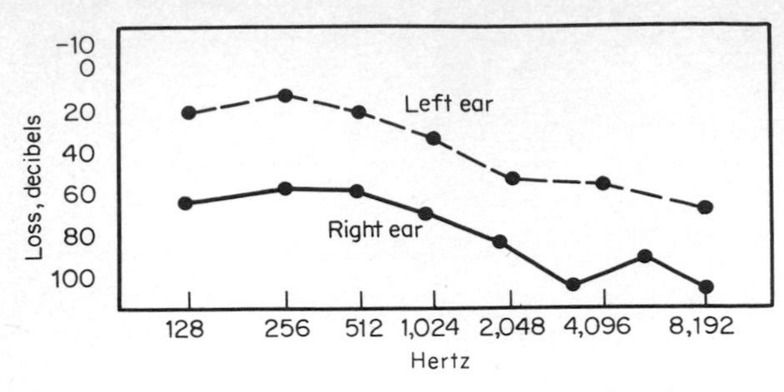

FIGURE D-8 Audiogram showing hearing loss

---

44. record

45. An audiogram with a hearing loss is shown in Figure D-8. The lower the line, the ______________ the hearing loss.

---

45. greater

46. There are different types of loss:
1. Middle ear or conductive loss (see Figure D-9). This is caused by an obstruction of some kind in the middle ear. The person suffering this type of hearing loss is not likely to be bothered much by loud noises and so will be less subject to fatigue from them.
2. Nerve deafness (see Figure D-10). Inner ear damage.
3. Acoustic trauma (see Figure D-11). This typically results from exposure to high noise levels. Note the sharp dip in the audiogram in the 3,000-to-4,000-hertz area, which is characteristic of acoustic trauma. As exposure continues, the notch opens wider, and eventually the person suffering from acoustic trauma can notice the loss of hearing.

Three kinds of hearing loss are: 1. ______________. 2. ______________. 3. ______________.

---

46. Conductive
Nerve
Trauma

47. How much loss constitutes a problem? Figure D-12 indicates what is serious loss. A person with more than a 25-decibel loss but less than a 40-decibel loss has a ______________ of hearing.

---

47. mild loss

48. Persons with certain types of hearing loss, such as nerve deafness, may be susceptible to more acoustic damage and should

---

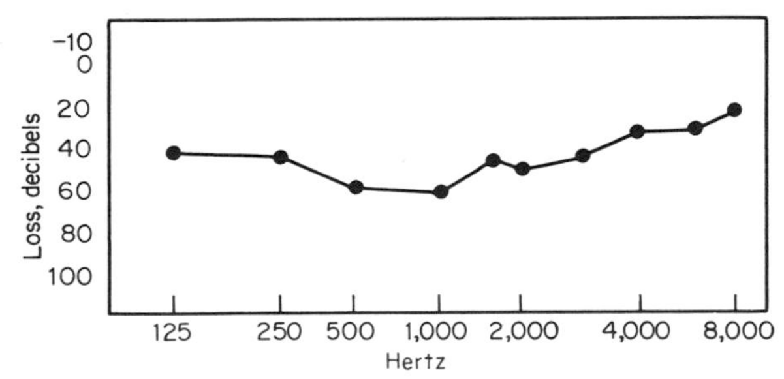

FIGURE D-9 Audiogram showing middle ear, or conductive, deafness

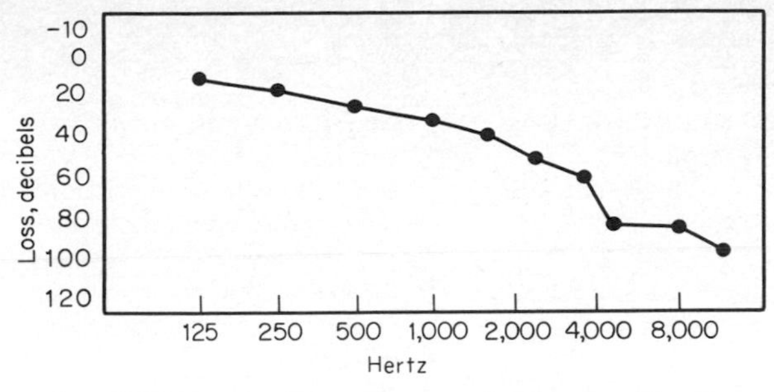

FIGURE D-10 Audiogram showing nerve deafness

not be considered for work in areas of high noise level. On the other hand, those with pure conductive deafness are not prone to additional loss and would be excellent in these areas. Both the type and degree of hearing loss should be considered in placement.

Both the ______ and the ______ of hearing loss should be evaluated when placing an employee on the job.

---

48. type
    degree

49. In addition to helping to place the employee, the audiometric examination provides a good permanent record. This record should become a permanent part of the medical record, should be kept confidential, and should not be removed from the permanent files. The audiogram will serve as a base upon which to evaluate subsequent changes in hearing acuity.

The audiogram is the permanent record of the level of ______ acuity at the time of hire.

---

49. hearing

50. It is important to remember that in this area the work done by the audiometric technician must be acceptable to medical and legal experts who will use the data.

The technician must be trained according to the principles of the American Academy of Ophthalmology and Otolaryngology (AAOO). The audiometer must meet the specifications of the American National Standards Institute (ANSI).

Technician training standards are set by the ______.
Audiometer specifications are set by the ______.

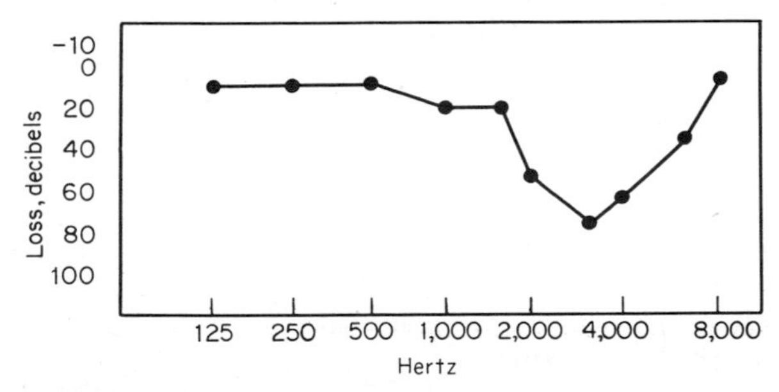

FIGURE D-11 Audiogram showing acoustic trauma

Class	*Name*	*Average loss in speech range*	*Remarks*
I	Normal	Less than 15 dB	
II	Near normal	More than 15 dB, but less than 25 dB	No difficulty conversing up to about 20 feet
III	Mild loss	More than 25 dB, but less than 40 dB	Difficulty with ordinary conversation when more than 5 feet away
IV	Moderate loss	More than 40 dB, but less than 65 dB	Difficulty with loud conversation when more than 5 feet away
V	Severe loss	More than 65 dB, but less than 75 dB	Difficulty with shouting when over 5 feet away
VI	Profound loss	More than 75 dB, but less than 85 dB	Difficulty with shouting at less than 5 feet
VII	Total loss	More than 85 dB	Loss of hearing of all speech

FIGURE D-12 Classification of hearing losses

50. AAOO
ANSI

51. The accuracy of an audiometer should be checked at least once a month. This check, known as a "biological calibration," consists of making audiograms for three or more young persons who have no history of previous ear disease or hearing loss. The testing conditions and technique used in the biological calibration must conform to current standards. At each test frequency the average threshold of the young persons should be within 5 decibels of the audiometer zero; if it is not, the acoustical calibration of the audiometer is probably incorrect.

If the audiometer is not in correct acoustical calibration, it should be returned to the manufacturer for service and adjustment. In any case, an audiometer should be returned to the manufacturer for service once a year.

Its accuracy should be checked at least ____________.

51. monthly

52. A hearing test room should be located in as quiet a place as possible, preferably within practical access but away from outside walls, elevators, heating and plumbing noises, etc. If the highest background noise levels do not exceed the values listed below, test-room noises will not affect test results:

Cycles	300 to 600	600 to 1,200	1,200 to 2,400	2,400 to 4,800	4,800 to 9,600
Decibel level	40	40	48	57	67

Acoustical treatment of audiometry rooms may prove to be a difficult task and should be undertaken only with the help of an acoustical consultant. Prefabricated audiometry booths usually are more satisfactory and may be cheaper than either reconstructed rooms or sound-treated rooms.

The important thing to remember is that hearing test rooms must be as ____________ as possible.

52. quiet

53. Hearing conservation medical control must include competent supervision by a physician to be considered adequate. This should include supervision of the audiometric facility to be certain that the audiogram data are accurate and reliable.

Supervision by a ____________ is mandatory.

---

53. physician

Section 6
NOISE: ENGINEERING CONTROLS

---

54. The methods of controlling noise for a particular situation may depend on many factors. No simple set of rules can be established to control noise.

For best results, one person might be appointed in an organization to coordinate the work on noise abatement.

One way to reduce noise is to control it at the source. An example of this would be to make sure that all bolts, screws, etc., are tight so that they will not rattle excessively.

Controlling noise at the ____________ is one method of noise reduction.

---

54. source

55. Substitution is a second method. An example of substitution is replacing metal tabletops with wood ones as a working surface for metal parts.

Two methods of control are:

____________ and

control at the ____________.

---

55. substitution
source

56. Isolation is a third method, An example of isolation is placing a particularly noisy operation in a separate room away from other operations, thus exposing only a few people to the noise instead of many.

Three methods of noise control are:

1. ____________.
2. ____________.
3. Control at the ____________.

---

56. Isolation
Substitution
source

57. One method of isolating noise is to construct a barrier or wall between the noise source and the ear.

The effectiveness of an isolating barrier is dependent on the material used for the barrier (its weight) and the frequency or type of noise being isolated.

For instance, a heavy concrete or brick wall will stop more sound than a plywood wall (see Figure D-13).

The ____________ the wall, the less noise will get through.

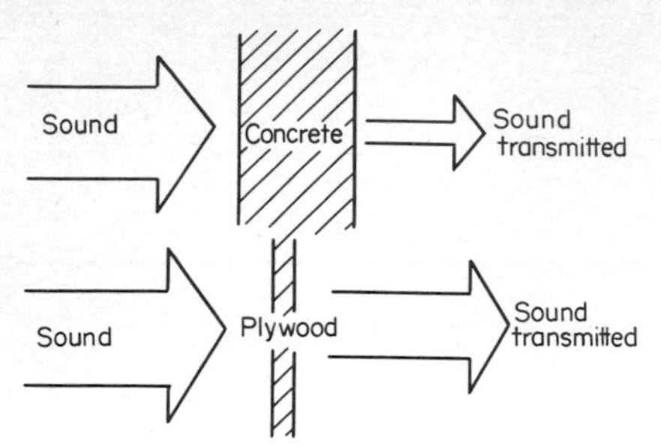

FIGURE D-13 Sound transmission

---

57. heavier

58. "Transmission loss" is a term describing the amount of sound stopped by the wall. It is ______________ with a concrete wall than with a plywood wall.

---

58. greater

59. Frequency is also a factor. High frequencies of noise are blocked more easily than low frequencies (see Figure D-14).

______________ (More, Less) high-frequency noise gets through a plywood wall than low-frequency noise.

---

59. Less

60. Since high-frequency noises are more damaging to the ear than low-frequency noises and since high frequencies are blocked more easily with mass, isolation is an effective noise-control method.

A fourth method of abatement is through the use of resilient mountings. If heavy machines are firmly bolted to concrete or wood floors, the floors are often transformed into huge sounding boards that not only amplify the noise but also spread it throughout the building. The use of rubber or other resilient mountings will usually reduce both noise and vibration. The use of resilient floor coverings is advisable to further reduce the noise level. Familiar examples are the spring mounting of electric refrigerators and the rubber mounting of automobile engines. In a plant, every machine will be a problem in itself, and thus the service of a technical expert is highly desirable.

Four methods of control are

1. Resilient ______________.
2. ______________.
3. ______________.
4. Control at the ______________.

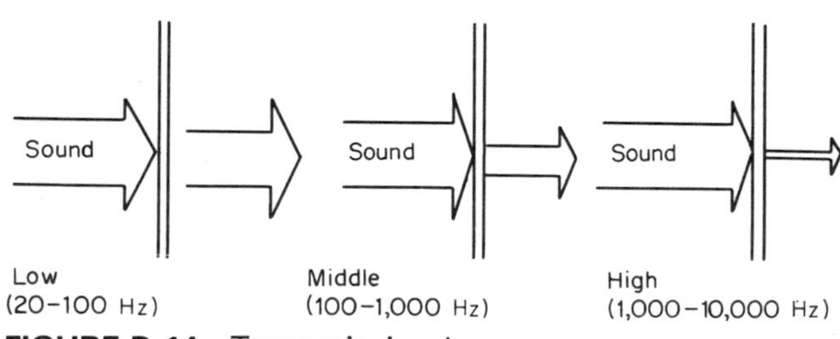

FIGURE D-14 Transmission loss

60. mountings
Isolation
Substitution
source

61. A fifth method is control of noise by use of sound-absorptive materials. Hard surfaces, such as brick and plaster walls, reflect sound and cause reverberation. If certain facing materials of known acoustical properties are applied to ceilings and walls, airborne noise will be partially absorbed. The more energy that is thus absorbed, the less that will be reflected back into the room to add to the original noise intensity. There are prefabricated acoustical tiles, acoustical plasters, sprayed-on compositions, and blankets which have been fabricated from very porous material, such as glass wool. Sound control by acoustical treatment is highly technical, and help should be sought from experts in the field. It must be remembered, too, that acoustical treatment can do nothing to lessen the direct noise of machines but is highly efficient in decreasing indirect or reflected noise.

The five methods of noise control are:

1. Use of ________________ materials.
2. ________________ mountings.
3. ________________.
4. ________________.
5. Control at the ________________.

---

61. sound-absorptive
Resilient
Isolation
Substitution
source

62. Each rise in the decibel level constitutes a great rise in sound pressures, as the formula is logarithmic. A doubling of sound intensity will raise the sound-pressure level only about 3 decibels.

Hence, if one machine creates 95 decibels, two machines will create not 190 but rather ________________ decibels.

---

62. 98

63. In the above example the noise level is exactly twice as loud at the ear, and yet it is only 3 decibels higher.

In this case, shutting off one machine will not cut the decibel level in half. It will reduce it by only about ________________ decibels.

---

63. 3

In effective noise control through engineering methods professional assistance is usually desirable, as this is a highly complex and technical area.

This section is meant to give only some small insight into the problem.

Section 7
NOISE: REDUCTION AT THE EAR

64. In some situations, even after sound-control measures have been applied, the noise level may still be too high because of the nature of the operation. For example, the operators of punch presses, drop hammers, chipping hammers, and the like cannot be protected by the noise-control measures already discussed. It is necessary in such instances that the operators wear properly designed and fitted ear protectors (earplugs), which will reduce the intensity of sound reaching the hearing mechanism. This is as necessary for the protection of ears as the use of safety goggles is for the protection of eyes.

The amount of sound protection offered by a good ear-protective device varies somewhat with design, but the device may be considered to reduce the level of the noise reaching the hearing mechanism by approximately 25 to 30 decibels.

In most cases a reduction of 25 to 30 decibels will reduce the noise to a ____________ level.

---

64. safe

65. Because ear protection is still not understood too well, a great deal of education is necessary to make hearing conservation programs effective.

The principal purpose of all ear-protective equipment is to reduce the intensity of harmful sounds before they reach the inner ear. Without this protection, the energy which is produced as a result of noise gradually destroys the delicate nerve endings which pick up sound vibrations.

Earplugs reduce the intensity of the ____________ at the ____________.

---

65. noise
ear

66. Experience in many plants has proved almost conclusively that properly fitted protectors, consistently worn, will prevent hearing damage where noise levels are hazardous. In one plant where a large number of workers were exposed to the same noise levels, one group used protectors, and the remainder did not use them. Later, audiometric measurements indicated that the protected group had relatively unchanged hearing ability, whereas there was noticeable lessening of hearing ability in the unprotected group.

____________ do work.

---

66. Earplugs

67. Generally, whenever workers are exposed regularly for many hours a day to noise levels above 85 decibels in the range of 300 to 600 or 600 to 1,200 hertz, the use of ear-protective devices is recommended.

____________ decibels will destroy hearing.

---

67. 85

68. Doctors are especially careful to advise workers with beginning or advanced hearing loss to wear ear-protective devices and thus prevent further deterioration of hearing due to noise. A worker *fitted properly* with effective ear protection can work in almost any noisy environment with no danger to hearing.

Ear protection must be properly ____________ to be effective.

---

68. fitted

69. The responsibility for hearing protection rests with management. In most plants, one individual should be given the task of following through with the details—probably the one who has charge of other protective equipment.

As in all safety programs ____________ is responsible for control.

---

69. management

70. There are many kinds of protective devices on the market. They can be divided into two general classes: (1) insert-type protectors and (2) earmuffs and helmets.

	The two types are: 1. ______________________ 2. ______________________
70. Plugs Muffs	71. The best ear protector is the one that is *worn.* Most protective devices seem to result in about the same amount of noise reduction, which is 20 to 35 decibels. When both helmet or muffs and insert-type protectors are worn, there is an additional 3- to 5-decibel total reduction of noise. The best ear protector is ______________________.
71. the one that is worn	72. Insert-type protectors are made of cotton impregnated with wax and of rubber or neoprene. In general, the material and shape of the insert-type devices have little to do with their effectiveness. Contrary to popular opinion, dry cotton affords little or no protection for excessive noise. The best ear protector is ______________________.
72. the one that is worn	73. Personal preference, as well as the degree of high-intensity noise, should govern the choice of muffs, helmet, or insert-type protection or a combination of them. Makers of ear-protective devices will undoubtedly be glad to send you samples of their products for you to examine. ______________________ should govern the choice of protection.
73. Personal choice	74. To be fully effective, insert-type protectors must be properly fitted and must be used at all times of exposure. Plugs must be ______________________.
74. properly fitted	75. Employees will be far more willing to wear protectors when they feel that they have been given ample opportunity to select the type best suited to them. The best way is to show several types of protection to the employee, explain their features, allow the employee to examine their construction and feel their softness, and emphasize that whatever device is chosen will be personally fitted. The best ear protector is ______________________.
75. the one that is worn	Supervisors can do a great deal in the course of their normal routine to promote hearing protection among employees. And the all-important thing is that employees who are *sold* on hearing protection are bound to spread some of their enthusiasm to their fellow workers.

END

The preceding sections should give the reader a good idea of what programmed instruction is. In most safety courses, the material is presented in the format shown in Sections 1 and 3. Literally thousands of programs are now being used on every conceivable subject. Anyone desiring a list of available programs can obtain it from the National Society for Programmed Instruction.

APPENDIX E

THE SAFETY GAME

In this appendix we shall discuss a new training technique, the management game, which simulates actual business situations in a classroom setting. The military service has long recognized the value of using simulations for educational purposes. War games, for example, are used extensively by the armed forces. The military has found that games are particularly valuable when it is impossible to participate in the actual situation and also where it would be very expensive to provide on-the-job experience. Games can be useful in business for the same purposes and reasons.

A business game consists of a controlled situation in which a team competes against intelligent adversaries and/or an environment to attain predetermined objectives. The environment and the controlling rules are designed to represent a business situation. Military games are composed of the same essential elements but represent military situations.

In a game the players contend with several interacting variables, some of which are under their control. There are definite relationships between the decisions made by the players and the results. The nature of these relationships is not fully known to the players but must be inferred from observations as the situations arise. The results and decisions of one period influence future conditions.

It is only recently that efforts have been directed toward applying the military gaming techniques to business problems, but rapid progress is now being made. A number of games have recently been developed by groups interested in executive training. The business games created thus far seem to be aimed mostly at management training. Their purpose is to give a member of management experience in the decision-making process, although the business games devised so far are mostly for the classroom situation.

The game presented here is intended to give safety professionals additional training in the decision-making processes that they must face in normal work. This game can be used in a classroom or individually. It can be self-administered, and it is perhaps the first of its nature in the field of safety.

The classroom version has been well tested over a period of years and has been found to be effective. I believe that this version of it will also be helpful to the individual student or reader who is genuinely interested in furthering his or her skills in accident prevention.

If you wish, use this game as a test of your comprehension of the material presented in this book or as a final examination in any course that has used portions of this book as a reference; it does attempt to cover most of the subjects important in safety management. If you accurately follow the instructions (really play the game), you may wish to compare your results with those of others who have played the classroom version of the game. Here is how they fared:

	Frequency rate	*Severity rate*	*Loss ratio*
Highest score	38.00	3,700	120
Lowest score	2.64	26	26
Average score	6.40	640	44

These scores were obtained in a controlled situation. If you wish to play the game somewhat as they did, give yourself a limited time in each section as follows:

Section 1: 1 hour

Section 2: 1½ hours

Section 3: 1½ hours

Section 4: 1 hour

And, of course, do not ever turn to the page containing the critique of the section until you have completed your recommendations to management.

PLAYER INSTRUCTIONS

You have been asked to analyze the safety program of a large manufacturer, the Eastabrook Manufacturing Company, located in your hometown. T. R. Eastabrook, treasurer of the company, has come to the conclusion that it is now time to take stock of the company's activities to ensure that it is doing all it can to retain its favorable insurance rating.

Mr. Eastabrook has opened all the necessary doors for you. You will contact all the appropriate operating executives of the company, learn about their safety problems, and observe their accident controls. Mr. Eastabrook is quite willing to reward you handsomely for your work and has agreed to pay you according to your success in this endeavor. He has agreed that your fee will be 50 percent of the difference between the loss ratio for the year and a 60 percent loss-ratio figure. Hence you can earn over $25,000 for your efforts.

Mr. Eastabrook has requested that you provide him with four interim reports containing your suggestions. As you submit these reports to him, he will provide you with the current frequency rate, severity rate, and loss ratio. These three indicators will, of course, reflect your suggestions. From the loss-ratio figure you can compute your earnings as the game progresses and at the end.

Now turn to Section 1 and read the descriptions of your contacts with the first four Eastabrook executives. When you have completed this section, you will be asked to identify problems and to submit your first report to Mr. Eastabrook. You will then compare your recommendations with the recommendations that actually would solve the problems of the Eastabrook Manufacturing Company.

If your recommendations compare favorably with those on the critique page, you will retain favorable frequency rates, severity rates, and loss ratios. If they do not, you will note all three indicators rising, as accidents are occurring in the plant.

Now begin your analysis.

SECTION 1

Contact 1: Company Treasurer

The treasurer of the Eastabrook Manufacturing Company is T. R. Eastabrook. Mr. Eastabrook has been with the company for 15 years, all this time as company treasurer.

Mr. Eastabrook provides you with the loss information shown in Figures E-1 and E-2. Since he considers you a safety expert, he has persuaded the president of the company to issue a directive stating that any of your suggestions will be agreed to. He is sure that the accident record will immediately reflect the effectiveness of the suggestions that you present.

Contact 2: Company President

The president of the company is P. R. Eastabrook. Mr. Eastabrook has been with the company for 30 years and for the last 10 years has been president. Before that he was vice president in charge of the central manufacturing division, superintendent of the central manufacturing division, and supervisor of the machine shop department.

Mr. Eastabrook appears to be very interested in safety in the plant. He freely admits that up until now he has paid no attention to it at all. He and the company treasurer, who are brothers, are vitally concerned about the present losses and definitely want them reduced. At the present time, they feel completely unable to handle this problem. They have pinned their hopes on you and are confident that you will help.

Mr. Eastabrook says that he will be glad to answer any questions about the company, its organization, its people, etc. He provides you with a company

EASTABROOK MFG. COMPANY
Loss Record

Year	Premium	# Acc.	# Comp. cases	Cost	Loss ratio	Exp. mod.
7 yrs ago	$76,504	280	53	$25,744	34%	11 CR
6 " "	70,453	270	56	53,626	76%	24CR
5 " "	67,913	181	35	28,998	43%	28CR
4 " "	78,527	175	38	15,935	20%	11CR
3 " "	67,570	252	29	39,045	58%	19CR
2 " "	62,971	371	33	50,367	80%	25CR
Last year	69,994	316	30	63,628	91%	25CR
This year	90,000					25CR

FIGURE E-1 Loss record, Eastabrook Manufacturing Company

EASTABROOK MFG. COMPANY
Injury report

Year	# Indust. first aid cases	# Disabl. inj.	Time charges	Man-hours worked	Freq rate	Sev. rate
7 yrs ago	23,960	78	630	2,583,000	30.1	243
6 " "	19,740	84	1,325	2,778,000	30.3	477
5 " "	19,840	50	712	2,829,000	17.1	252
4 " "	18,497	48	390	2,649,000	18.1	147
3 " "	20,000	38	962	2,508,000	15.1	384
2 " "	22,480	38	1,240	2,517,000	15.1	493
Last year	19,430	35	1,575	2,796,000	12.5	564
This year				3,000,000		

FIGURE E-2 Injury report, Eastabrook Manufacturing Company

organization chart (Figure E-3) and a sheet showing the layout of the company's plants. He further states that the organization chart actually shows how the company does operate. Mr. Eastabrook suggests strongly that you contact each important person shown on the organization chart and that you make thorough surveys and submit recommendations where needed. He would like to be kept informed but agrees that your reports should go to his brother, who can keep him up-to-date, He has already issued orders to everyone in the plant to go along with any reasonable recommendations that you submit.

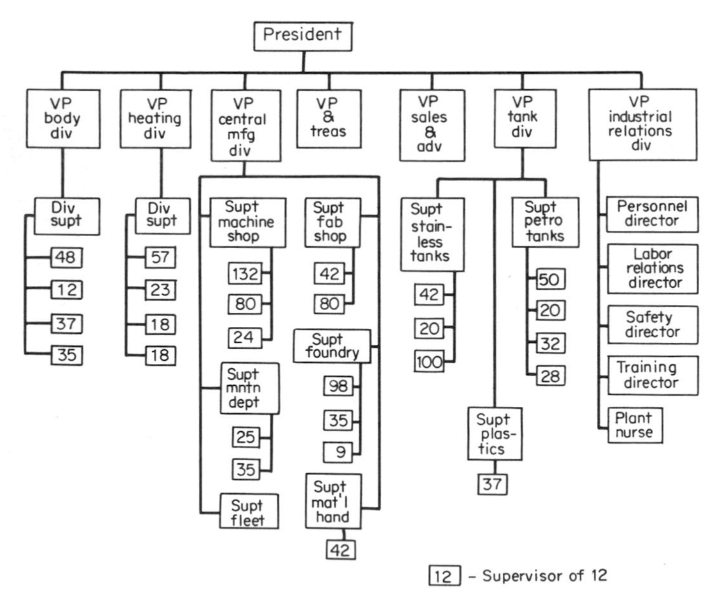

FIGURE E-3 Organization chart, Eastabrook Manufacturing Company

Upon your questioning, Mr. Eastabrook provides the following information: In the past, neither he nor any other executive has participated in the safety program at all. He has officially delegated responsibility for safety to nobody in particular, but he believes that the supervisors throughout the plant probably know that they are responsible, as the safety director does. No job descriptions for the safety director or anyone else in the company have been made up. In Mr. Eastabrook's opinion, the duties of the safety director would probably be to inspect, to keep accident records, to keep management informed, and to tell the supervisors of the various departments "what to do."

Up until this point, Mr. Eastabrook has been too involved in other company operations to know exactly what the safety program was, what the safety director's job was, etc., and he suggests that you contact the safety director himself for the details of the program. He provides the following information on the Eastabrook Manufacturing Company. There are four divisions in the company:

BODY DIVISION. Engaged in the manufacture of dump-truck bodies. Angle and plate steel is received, cut to size, jigged, welded into the body framework, sandblasted in a large cabinet, and spray-painted. This is done on one long, large assembly line.

HEATING DIVISION. Engaged in the manufacture of residential gas-fired hot-air furnaces. Formed sheet steel and all parts are received from the central manufacturing division, and the units are assembled, painted, and packaged for delivery to dealers.

TANK DIVISION. Engaged in the manufacture of tank-truck bodies for petroleum, milk, and other industries utilizing bulk liquid hauling equipment. This division is composed of three separate departments:

1. Stainless tanks. Manufacture of stainless steel tanks. Sheet steel, insulation, and structural steel are received, formed, and assembled on large assembly lines.

2. Petroleum tanks. Manufacture of bulk petroleum product haulers. Sheet metal and structural steel are received, formed, and assembled.

3. Plastic tanks. Manufacture of *(a)* plastic tank trucks and *(b)* pilot operations in the use of plastics for insulating material and a number of other products. The plastic tanks are fabricated of fiber glass.

CENTRAL MANUFACTURING DIVISION. Engaged in the manufacture of subassemblies for the other divisions. In addition, it puts out several products of its own, as well as job work for other industries. This division is composed of the following departments:

1. Machine shop—large machining facilities

2. Foundry—gray-iron foundry producing a variety of castings

3. Fabrication department—engaged in the manufacture of a variety of steel-fabricated products much on a job basis
4. Maintenance department—for the entire company
5. Material handling department—in charge of all receiving, handling of work in process, and shipping for the entire company.
6. Fleet department—engaged in trucking of products throughout the United States

Mr. Eastabrook also provides you with the plant layout (Figure E-4).

Contact 3: Vice President, Industrial Relations

The industrial relations vice president of the company is Irene Radtke. Ms. Radtke has been with the company for 32 years and has been a vice president for 10 years; before that she was personnel manager for 5 years, safety director for 10 years, and a shopworker for 7 years.

Upon your questioning, Ms. Radtke provides you with the following information: She feels that the present safety program is quite weak in the inspection area. She states that she used to inspect the entire plant weekly and had a system of paperwork set up which effectively got things done. Her safety

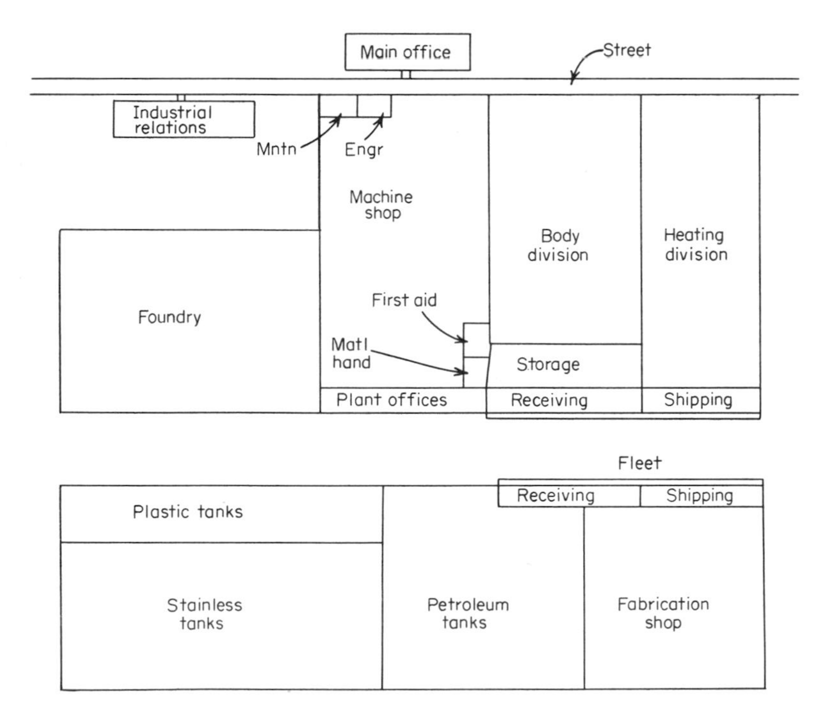

FIGURE E-4 Plant layout, Eastabrook Manufacturing Company

committees, which she originally set up and which are still in operation, used to religiously inspect their various divisions. She feels that the superintendents' safety committee, which she also set up, accomplished a great deal in safety. However, in the past few years, she herself has not had too much time to devote to safety because of her various labor relations problems. She would like your ideas on an internal inspection program and promises that she will agree to any reasonable suggestion in this area.

Ms. Radtke also feels that the company's present training of employees is pretty bad. Upon your further questioning, she tells you that there is no safety indoctrination for new employees and that, after being hired, they go directly to the supervisor, who assigns them to an experienced employee for on-the-job training. This usually lasts for a day or two. There are no written job instructions on any job throughout the plant. As far as she knows, safety is not necessarily included in this on-the-job training, at least not to any great extent. For those employees transferring from one department to another, no particular training is required. In fact, up until now, she had not even considered this possibility.

Ms. Radtke feels that the training of supervisors is quite adequate at this point. There is no training program for the newly appointed supervisor, but the company does make considerable use of management meetings at the top level and at the division level. The company believes in meetings as an effective communications medium and training device, and they are used a good deal for further training of all supervisors throughout the company. The meetings she refers to are not necessarily confined to safety. In fact, they are set up primarily as production meetings. However, safety is discussed in the meetings periodically. But safety is regularly included in the meetings of the superintendents' safety committee; in fact, it is the main reason for these meetings.

There are no written rules, precepts, or standard operating procedures in the company. Generally, however, Ms. Radtke feels that supervision is doing a good job without any written rules. The supervisors all know what is safe and what is not. She does feel that the supervisor, as a rule, sets the correct example. She does not believe in the use of safety manuals and has never made up any. When you ask whether safety is included as an integral part of all production, Ms. Radtke says that basically it is not and never has been. It is usually treated as a separate entity.

Contact 4: Safety Director

The company's safety director is Sam Dowling. Mr. Dowling is an old-timer in the plant but has been involved in safety work for only about one year. Before being assigned to his present job, he spent four years as a personnel assistant. In this capacity, he ran the company's suggestion plan and the company publication. Before being transferred to the personnel department, he was supervisor in the heating division.

Mr. Dowling, at your request, describes the safety program as follows:

Contests are used as often as possible throughout the plant. Many varieties have been used during the past several years. At present, a bingo contest is in effect. Once each day the plant nurse draws a number. Each department has a bingo card posted on the main bulletin board. If a lost-time accident occurs, the entire department is eliminated from the game. The first department to get a bingo then has a drawing for a prize. Mr. Dowling feels that this contest is doing a beautiful job and probably will bring the loss ratio right back in line.

In addition, Mr. Dowling has been stepping up his bulletin-board campaign in the last six months. He has added new bulletin boards throughout the plant and has purchased a number of new posters from the National Safety Council, which he is using at a rapid pace. He recently started safety film showings in the cafeteria at lunchtime. He gets a number of films from the local library and also orders them from the National Safety Council. He feels that this program is particularly effective as he reaches 15 or 20 employees every week and is sure this will reduce the accident frequency rate. He still publishes the company magazine and has stepped up his safety messages in this a great deal.

When queried about the rest of the safety program, Mr. Dowling states that he does periodically inspect the plant. He uses no particular time schedules for making his inspections and could not definitely pinpoint the last time that an inspection was made. On these inspections, when he has a suggestion to make, he goes directly to the employee involved and never has any trouble getting the mistake corrected. He keeps no records of any of these inspections. If the suggestion involves the maintenance department regarding the construction of a guard, for example, he usually phones the department directly and asks for someone to come and take care of the matter. He has never made any particular follow-ups on these matters other than checking on them when he is in the area.

Mr. Dowling also states that the company has a number of safety committees in operation as follows:

1. Each division is supposed to have a safety committee composed of two supervisors and four workers, which is supposed to meet monthly for a division inspection. However, only one division is actually still doing this. Most of the others have not made any surveys or had any committee meetings for at least six months. No records are kept on any of these committees.

2. There is a superintendents' safety committee, and Mr. Dowling feels that this is going along quite well. From his file, he pulls out the bylaws of this organization and reviews them for you. The members of this committee meet monthly, review the accident record, discuss accident problems, and usually have some outsider come in and talk to them. They are supposed to be setting policies for the company. Every other meeting is a luncheon meeting, held away from the plant. At this type of meeting, no safety discussion is held. Again, no particular records are being kept of this comm ttee's activities.

When queried about any special inspections that are made in the plant, Mr. Dowling mentioned that the boilers and the plant elevators are inspected by the insurance company. Other than this, he could think of no other special inspections being made.

As far as Mr. Dowling could recall, this pretty well summarizes what he feels is the safety program of this company. He also shows you his accident record (see Figure E-2).

Mr. Dowling states: "Look at that frequency rate—you can't argue with success!"

At this point, *stop* reading and review your discussion with the four people contacted. Identify any problem areas you perceive at the Eastabrook Manufacturing Company.

Now, briefly write your recommendations to Mr. Eastabrook.

Do Not Turn Page Until Completed

CRITIQUE: For the first four contacts you should have recommended these things: If you did not, charge yourself these costs:

	No. of Accidents	Time Charges	Cost
CONTACT 1: *Treasurer*			
CONTACT 2: *President*			
A written management policy	12	250	$6,500
Definite procedures which fix accountability	12	250	6,500
CONTACT 3: *Vice president, industrial relations*			
A training program for employees (orientation)	10	200	5,000
A supervisory training program	10	200	5,000
CONTACT 4: *Safety director*			
An inspection system	10	200	5,000
Total charges above	0	0	$938
Normal charges (add)			
TOTAL			
	×1.3	×1.3	÷$22,500
Frequency rate	______		
Severity rate		______	
Loss ratio			______

SECTION 2

Contact 5: Company Personnel Director

The personnel director of the organization is Pat Demeter. Mrs. Demeter has been the personnel director of the company for the past 10 years. Before that, she was a personnel assistant. She started with the company 12 years ago on the personnel assistant job.

Upon your questioning, Mrs. Demeter gives you the following information: Most employees are obtained through newspaper advertising. Each department requests the new employees it needs, and Mrs. Demeter places ads in the newspaper for the jobs. Each applicant is required to fill our an application blank and is interviewed by Mrs. Demeter. She usually makes a telephone reference check of the past two employers listed on the application blank. The applicant is then required to undergo a physical examination, which may be performed by any one of four different local physicians. As Mrs. Demeter describes it, the physical examination given seems adequate. Mrs. Demeter has never met the physicians involved and has discussed problems with them only over the phone.

The applicant is next given aptitude, mechanical ability, and interest tests. Mrs. Demeter keeps a file on each person with the results of all the tests given, including the physical. She then decides whether the applicant should be hired. Once hired, the new employee is sent directly to the supervisor, who usually places the new person with a more experienced worker for training for at least one full day. After 30 days' probation, the supervisor decides whether to keep the new employee. There is no job analysis program, nor are there any job physical requirements specified throughout the company.

An applicant for foundry work is also given an x-ray, and the film is sent to a doctor. Mrs. Demeter stated that every foundry employee is reexamined as specified by the doctor, usually on a two-year recheck basis. The company is approximately 15 months or 100 people overdue on these reexaminations. Foundry employees are the only employees given reexaminations.

Contact 6: Plant Nurse

The plant nurse is Miss Regina Nuefeldt, who has been the nurse for the company for 15 years. In your discussion with Miss Nuefeldt, you learn the following information about the company health program: The first-aid facilities that the company provides are adequate and are available to all shifts of the plant and to both plants. All first-aid attendants are adequately qualified for their jobs. There is no medical direction to the program. There are no catastrophe or disaster plans. Facilities are available to transfer any injured employee to

EASTABROOK MFG. COMPANY • THREE-YEAR ACCIDENT ANALYSIS

Portion of body	Last year			Two years ago			Three years ago		
	% doctor cases	% LT acc.	% cost	% doctor cases	% LT acc.	% cost	% doctor cases	% LT acc	% cost
Eyes	25	-	-	22	-	25	41	-	10
Head and face	9	14	2	9	12	2	6	12	2
Trunk	6	11	5	9	11	15	6	17	10
Arms	13	18	10	13	20	10	9	18	22
Hands and fingers	34	35	19	30	32	32	29	24	39
Legs	7	11	2	7	11	10	1	7	8
Feet	5	11	1	9	14	5	5	12	5
Toes	1	-	1	1	-	1	3	10	4
Whole body	-	-	60	-	-	-	-	-	-
Accident type									
Machinery		9	15		3	29		5	20
Material handling		28	10		29	10		55	39
Hand tools		13	1		-	2		3	5
Falling objects		13	2		18	12		15	5
Slips and falls		16	8		21	10		7	12
Bumping objects		15	3		15	10		5	2
Chemicals		3	1		2	1		5	2
Flying particles		-	-		-	25		-	10
Miscellaneous		3	60		12	1		5	5
Injury type									
Amputations		6	15		3	20		10	10
Fractures		22	8		29	10		34	10
Lacerations		22	2		25	10		17	37
Foreign body in eye		6	-		-	25		2	10
Contusions		25	10		18	10		27	23
Strains		6	2		14	15		10	18
Burns		6	2		7	2		-	2
Dermatitis		3	1		-	3		-	-
Miscellaneous		4	60		4	5		-	-

FIGURE E-5

the hospital. The nurse has no directory of qualified physicians posted in her room, as she knows which ones to call for which type of injury.

At the present time there is no off-the-job health program in effect. The company is looking for a new nurse for plant number 2, as the present one will be quitting next month, but it does not know where to obtain one.

A review of the records system which the plant uses shows that the nurse keeps the daily log of treatments. She summarizes this monthly and sends it to the vice president of industrial relations, the safety director, and the insurance carrier. She does not know what happens to those copies after these people have seen them. She makes up an annual report also and, at your request, provides you with her report and her analyses of accidents that occurred during the last three years (Figure E-5).

While still discussing the accident record system, you obtain the following information from the nurse and the plant safety director: The former safety director used to investigate all accidents, lost time, doctor cases, and near misses. This has not been done for five years, and no accident investigation reports have been filled out since then. Mr. Dowling feels that the frequency rate is coming down without these time-consuming investigations and that they obviously are not needed anymore. Also, he no longer has the time to investigate these accidents. Miss Nuefeldt feels that she gets enough information without investigations, and she shows you her analyses (Figures E-6 to E-8).

EASTABROOK MFG. COMPANY • ANNUAL ACCIDENT REPORT

Department	Man-hours	First aid cases	Doctor cases	Disability injury	Time charges	Cost
Body Division	214,000	1,750	28	3	142	1,679
Heating Division	232,000	1,362	22	2	112	1,308
Tank Division						
Stainless	164,000	972	16	2	79	932
Petroleum	260,000	1,943	32	4	158	1,862
Plastic	38,000	678	13	1	63	746
Central Mfg.Div.						
Machine Shop	422,000	3,110	51	6	252	2,982
Foundry	242,000	3,882	63	7	315	43,669
Fab. Shop	244,000	2,240	35	4	173	2,050
Maintenance	120,000	1,557	25	3	126	1,492
Fleet	152,000	584	9	1	45	5,595
Mat'l.Handling	210,000	1,158	19	2	94	1,120
Office & Misc.	498,000	194	3	0	16	186
Totals	2,796,000	19,430	316	35	1,575	63,621

Freq. rate for year
$$\frac{35 \times 1{,}000{,}000}{2{,}796{,}000} = 12.6$$

Sev. rate for year
$$\frac{1{,}575 \times 1{,}000{,}000}{2{,}796{,}000} = 564$$

Loss ratio for year
$$\frac{\$63{,}621}{\$69{,}994} = 91\%$$

FIGURE E-6 Annual accident report, last year

EASTABROOK MFG. COMPANY • ANNUAL ACCIDENT REPORT

Department	Man-hours	First aid cases	Doctor cases	Disabiling injury	Time charges	Cost
Body Division	143,000	2,700	45	5	149	2,440
Heating Division	211,000	900	15	2	50	815
Tank Division						
Stainless	164,000	1,575	26	3	87	1,425
Petroleum	260,000	2,022	33	3	113	1,832
Plastic	51,000	674	11	1	47	610
Central Mfg. Div.						
Machine Shop	372,000	3,595	59	6	198	23,259
Foundry	192,000	4,045	67	7	224	3,662
Fab. Shop	144,000	2,920	48	5	163	12,642
Maintenance	120,000	1,800	30	3	99	1,630
Fleet	152,000	449	8	1	25	407
Mat'l. Handling	210,000	1,350	22	1	60	1,222
Office & Misc.	498,000	450	7	1	25	423
Totals	2,517,000	22,480	371	38	1,240	50,367

Freq. rate for year $\frac{38 \times 1,000,000}{2,517,000}$ = 15.1

Sev. rate for year $\frac{1,240 \times 1,000,000}{2,517,000}$ = 493

Loss ratio for year $\frac{\$50,367}{\$62,971}$ = 80%

FIGURE E-7 Annual accident report, two years ago

Contact 7: Vice President, Tank Division

The vice president of the tank division is T. A. Nelsen. Mr. Nelsen has been with the company for 25 years; he has been a vice president of this division for 5 years, and he came up through the ranks. He is very cordial and friendly, and he offers to take you through his division himself. As the survey is made, Mr. Nelsen explains the various processes.

Angle and plate steel is received from the receiving department and is cut to size on large shears and presses. Steel is jigged in specially designed permanent metal jigs, and the body framework is fabricated. At this point, the framework is transported to the stainless department, the petroleum department, or the plastic assembly line. Up to this point, the operation is very similar to that in the body division, and the hazards are similar. Controls appear to be generally quite good in this first part of this division.

MACHINERY. Guarding on shears is adequate, and on presses adjustable barrier guards are used where individual die guards have not yet been built. There is a program of providing individual die guards on all dies, and within several more years this program will be completed.

HANDLING. All handling of materials is under the jurisdiction of the central manufacturing division.

HOUSEKEEPING. Housekeeping is generally good. Aisles are well marked,

EASTABROOK MFG. COMPANY • ANNUAL ACCIDENT REPORT

Department	Man-hours	First aid cases	Doctor cases	Disability injury	Time charges	Cost
Body Division	150,000	1,200	15	2	58	1,256
Heating Division	200,000	1,600	20	3	77	1,620
Tank Division						
Stainless	233,000	1,800	23	3	87	1,710
Petroleum	250,000	2,000	25	4	96	1,905
Plastic	5,000	50	0	0	0	0
Central Mfg. Div.						
Machine Shop	470,000	5,000	63	10	241	24,574
Foundry	180,000	3,400	43	6	164	3,220
Fab. Shop	120,000	1,000	13	2	48	950
Maintenance	100,000	750	10	2	38	760
Fleet	150,000	1,200	15	2	58	1,140
Mat'l. Handling	200,000	1,600	20	3	77	1,520
Office & Misc.	450,000	400	5	1	18	380
Totals	2,508,000	20,000	252	38	962	39,045

Freq. rate for year

$\frac{38 \times 1{,}000{,}000}{2{,}508{,}000} = 15.2$

Sev. rate for year

$\frac{962 \times 1{,}000{,}000}{2{,}508{,}000} = 384$

Loss ratio for year

$\frac{\$39{,}045}{\$67{,}570} = 58\%$

FIGURE E-8 Annual accident report, three years ago

work in process is stored in specified areas, and few problems are apparent here.

WELDING. Specially designed large weld frames are used. All welding is gas, and proper protective equipment is used by the welders. All welds are magnafluxed.

SANDBLASTING. The bodies, once constructed, are sandblasted in their entirety in large especially designed blast rooms. Nonsilica abrasive grit is used. The sandblast cabinet is built right into the production line to facilitate the handling of bodies. The blaster seems to use adequate and proper protective equipment. Hand grinding is performed on most welds in addition to the sandblasting. In the grinding area, pneumatic 5-inch hand grinders are in use.

PAINTING. After sandblasting, the bodies are conveyed to a paint-spray booth. This is a specially designed water-wash paint-spray booth, which appears to be very adequately vented. There is little overspray noticeable. Lead-base paints are used for the painting, and the operator is provided with, and seems to be wearing, a respirator. The respirator is approved by the Bureau of Mines and appears to fit well. After being painted, the subassemblies go through a drying oven. This appears to be a standard procedure and presents no unusual problem.

STAINLESS DEPARTMENT. This department receives the body frames and constructs the tanks. Stainless steel is received in sheets and shaped on the inside of the body frame, forming the interior of the tank. Then cork insulation is applied over the entire surface area. Next, stainless sheets are fastened onto the exterior of the frame, forming the exterior of the tank. Endpieces, both circular and elliptical, are fabricated—the circular endpieces are spun and then attached to the ends of the tank. All pieces are welded in place. All welded joints, both inside and out, are hand-ground with 5-inch pneumatic hand grinders; the inside grinding is done within the enclosed tank.

Much work is done from movable steps—up to 12 feet high. These are wooden stairways that can be moved from place to place. During the survey, you note that no protection is provided for those going up the stairs or standing at the top. Much work is performed at the top of the steps.

PETROLEUM DEPARTMENT. The operations in this department are very similar to those in the stainless department. The body frames are received, and the tanks are constructed in large assembly lines. Plate steel is formed and welded to the interior of the frames. The tanks are only single-shell tanks, as no insulation is to be applied. All of them are welded. However, in this type of tank the joints do not have to be ground. With the exception of not providing insulation, grinding joints, or applying the outer shell, the operations are the same as in the stainless department. General controls in this department seem to be quite good.

PLASTIC DEPARTMENT. This department is only four years old and is still in its infancy, according to Mr. Nelsen. Plastic tank truck manufacturing is a little past the pilot stage, and a small production assembly line is now in operation. Tanks are fabricated of reinforced plastic. Operations include vacuum molding, compression molding, and hand lay-up operations. Polyester resin in a styrene monomer catalyzed and benzoyl peroxide and dimethyl phthallate is the basic system used. Reinforcing material is fiber glass.

After the tanks are formed, they are sent to a separate area and lifted on, and fastened to, the truck bodies. In addition, the plastic department is piloting an operation of pouring polyurethane resin in place on some of its tanks. This may be used in the future as an insulating material. This is a polyurethane foam operation in which stypol resin (a prepolymer of the polyalcohol) and toluene diisocyanate plus catalyst are used. The system is genetron-blown, and pouring lasts for about 30 seconds. An exhaust blower is used in the area to remove fumes. All employees handling the resin wear protective gloves and clothing.

Contact 8: Vice President, Heating Division

The vice president of the heating division is Howard Dunn. Mr. Dunn meets you and immediately refers you to Harvey Heathcote, division superintendent. Mr. Heathcote grudgingly agrees to take you on a survey of the operations of the heating division. The division, Mr. Heathcote explains, produces residential gas-fired hot-air furnaces under the Xerxes trade name and under several other names for national retailers. Formed sheet steel and all parts and subassemblies are received from the central manufacturing division and subcontractors, and the heating division's prime job is an assembly operation.

During the tour of the division, you observe the following:

1. Receiving. All materials are received by forklift truck from the various sources. All material handlers and material handling equipment are under the jurisdiction of the central manufacturing division.
2. No unusual housekeeping problems are noted during your survey—several untidy spots are observed and presented to Mr. Heathcote, who says he will correct them when he gets a chance.
3. Some machine-shop type of machinery is in use in the department, and many power hand tools are in use in the assembly operation. Guarding of the general machinery seems to be adequate. All hand tools used are electric. Maintenance on the hand tools is performed only on breakdown. There is a two-pronged electric system here, as the plant is comparatively old.
4. Large, long assembly and subassembly lines are set up—the operation is almost totally conveyorized.
5. All assembled sheet-metal work is spray-painted on a conveyorized system. The spray-painting booth is a part of the total conveyor assembly line, is water-wash, and appears to be well ventilated. The drying oven appears to be a standard installation. Respirators are provided for the spray painters when lead-base paints are used.
6. Shipping. All units are crated and then sent to shipping, A number of circular saws are used in this crating department, and saw guards are in evidence on some of the saws.
7. Use of personal protective equipment in this department is basically good. There is almost a 100 percent use of eye protection, and most employees engaged in manually handling heavy materials are wearing foot protection.

At this point *stop* reading and review your discussion with the four people contacted. Identify any problem areas you perceive at the company.

Now, briefly write your recommendations to Mr. Eastabrook.

CRITIQUE: For the second four contacts you should have recommended these things:	If you did not, charge yourself these costs:		
	No of Accidents	Time Charges	Costs
CONTACT 5: *Personnel director*			
Ensure that the physician is familiar with the plant operations	5	100	$2,500
Ensure that information from the physician is actually used in placement	5	100	2,500
Have all foundry x-rays brought up-to-date	1	1,000	10,000
CONTACT 6: *Plant nurse*			
An accident investigation procedure	10	200	5,000
Specialized assistance should be brought in to help evaluate this area	10	200	5,000
CONTACT 7: *Vice president, tank division*			
Noise controls, which include *all* these: *a.* Audiometric examinations for all new lines; *b.* Noise-abatement plans; *c.* An ear-protection plan	1	600	6,000
An oxygen check of the tank	1	1,500	15,000
Handrails be constructed on the platforms	1	250	2,500
Specialized industrial hygiene assistance be brought into the plastic department	1	1,500	15,000
CONTACT 8: *Vice president, heating division*			
Electric grounds be provided	1	1,500	15,000
The saw be provided with a guard	1	250	2,500
Total charges: Section 2			
Total charges: Section 1			
Normal charges (add)	0	0	$1,875
TOTAL			
	×0.67	×0.67	÷$45,000
Frequency rate	______		
Severity rate		______	
Loss ratio			______

SECTION 3

Contact 9: Vice President, Central Manufacturing Division

The vice president of the central manufacturing division of the company is Charles McDonald. Mr. McDonald has been with the company for 38 years and has come up through the division. He hastily explains to you that the surveys of the various departments of his division should be made with his various department superintendents. He calls them individually on the phone and tells them that you will be coming to see them.

Mr. McDonald quickly goes over the operation of his division. His is the largest division and is pretty much the heart of the organization. The other divisions depend on central manufacturing for parts of their product, for materials handling, etc. The departments under the central manufacturing division are the following.

MACHINE SHOP. This is one of the largest machine shops in the area and has adequate machinery to handle almost any job presented to it. It is basically one large job shop, producing pieces for the various departments, as well as job work for outside industry. The machine shop does not really seek this job shopwork, but gets it because of the department's reputation for exceptional work and also because of the wide range of plant facilities available.

FOUNDRY. This is a gray-iron foundry, producing castings for the other divisions and also a large amount of job shopwork. Castings range from very small up to several tons of metal cast.

FABRICATION SHOP. This is mainly a steel fabrication job shop. This department has plant facilities and equipment to handle almost any job, and it uses these resources in jobbing for other industries. The department now manufactures and produces any type of structural steel work to customer design.

MAINTENANCE DEPARTMENT. This department has the responsibility for the maintenance of all equipment for all the divisions. It has a staff of machine maintenance personnel, electricians, carpenters, etc.

MATERIAL HANDLING DEPARTMENT. This department is responsible for all material handling functions for all divisions, including shipping, receiving, storage, and handling. Forklift trucks, overhead crane operators, and hitchers are all in this department. Employees of this department work throughout both the company's plants.

FLEET DEPARTMENT. This is actually a trucking company. Tractor-trailers of the department ship the finished products all over the United States.

As far as safety is concerned, Mr. McDonald says that he has never paid too much attention to the prevention of accidents in the past. He has been concerned with eye accidents of late, however, and mentions that in recent years several serious eye injuries have occurred in the machine-shop area. He states that he will agree pretty much to whatever suggestions you may have for his division. Whenever he is in the various departments of his division, he notices that only about one-third of the employees are actually wearing eye protection, although protection has been provided to every employee.

With the exception of the eye-protection problem, Mr. McDonald feels that his division is doing a good job in safety. He does not know whether the record will bear him out on this because he never gets to see the record. He does remember a silicosis case last year, but he cannot recall any of the details of this and suggests that probably it was a case that they should not have had to pay for, although he does not know for sure.

Contact 10: Foundry Superintendent

The superintendent of the foundry department of the central manufacturing division is Frank Dryden. Mr. Dryden has been superintendent of the foundry for 20 years; before that he was cleaning room foreman.

A survey of the foundry and a discussion with Mr. Dryden reveal the following information: Sand conditioning in this foundry is an automatic process. The sand is returned from shake-out automatically, by conveyor, and one operator controls the mixing of the old sand with the new sand, oil, etc. This is all done by push button. The clean sand is then returned to the molding station by overhead conveyor. Core sand is also mixed automatically and is conveyorized to the core room for use. The core room is a separate room from the rest of the foundry. The core ovens are adequately ventilated. Three core washes are used: Jones Ferro-Kote, #17-65, and Jones Griptite. These core washes are sprayed on wet. The core-wash spraying area is cleaned about every month.

Molding is conducted separately from the shake-out station. Silica-containing parting compounds are used. The sand slinger is used for floor molding, and no respirator is provided for the sand-slinger operator. Several shake-out stations are in use. All are provided with local exhaust ventilation which appears to be adequate. Respirators are provided to the shake-out operators. However, at the time of this survey they were not being used.

All casting cleaning is done in an isolated room away from the rest of the foundry. All small castings are precleaned in a Rotoblast, which is well ventilated and in good condition. A large number of tumbling barrels are employed which are all adequately ventilated. Stand grinders, hand grinders, and swing grinders are all used. All appear to be ventilated as much as they can be. All ventilation is attached to a dust collection system. The bags of this dust collection system are rapped every other week.

Housekeeping throughout the foundry is generally good. Use of personal protection is also generally good. Safety shoes are worn by almost 100 percent of the employees, and eye protection is mandatory in this area. Foundry gaiters, aprons, gloves, etc., are all provided and used when necessary.

Contact 11: Machine Shop Superintendent

The superintendent of the machine shop of the central manufacturing division is Max Shinners. Mr. Shinners has been superintendent of the machine shop for one year. Before that, he was in the production control department of the plant.

A survey of the machine shop and a discussion with Mr. Shinners reveal the following information: The layout of the machine shop is classed as a product layout because so much of the work done here can be considered job shopwork.

1. Lathe area. Several engine lathes, a number of turret lathes, and a few automatic screw machines are in this area. Transmission guarding is generally very good. However, your survey did reveal four belt and pulley guards missing on turret lathes numbers 12, 147, 138, and 204. The automatic screw machines are not provided with lined tubes.

2. Drills. A number of single- and multiple-spindle drills are in use. Three-fourths of them are operated by women; none of the operators are provided with or are wearing hair protection.
3. Milling machines. The 12 to 15 milling machines in this area are provided with good transmission guarding.
4. Saws. All guarding of metal saws appears to be satisfactory.
5. Planers. All planers seem to be adequately guarded with barrier guard-rails around the ends of the machines.
6. Shapers. There appears to be no problem here.
7. Grinders. Many grinders of different types are in use in this department. All storage of grinding wheels is in vertical racks, and humidity is controlled. The use of flanges, from your observation, appears to be correct. When the toolroom hands out a grinding wheel, it specifies to the operator the speed at which the wheel should be used. After this, it is up to the supervisor to ascertain whether this speed is actually being held to. In a survey of the grinding area, several things are apparent. On approximately half of the stand grinders, you notice that the tool rests are adjusted about 1 inch away from the grinding wheels. All grinders are provided with cast-iron guards. All wheels are either 2 or 4 inches in thickness, all guards are ⅝ inch thick, and the wheels are 6 to 24 inches in diameter.

Housekeeping throughout the department appears to be quite good. Use of personal protective equipment is noted as follows: About 50 percent of the employees are wearing eye protection, about 50 percent are wearing safety shoes, and no ear protection is noted.

Contact 12: Superintendent of the Fabrication Shop

The superintendent of the fabrication shop is Frank Schildkrant. He has been superintendent for five years and came up through the fabrication-shop ranks. The fabrication shop is engaged in the production of a variety of fabricated articles all made out of steel. The department has many large-capacity shears and presses plus plant facilities and material handling equipment available at few other companies. Consequently, a number of outside industries have given the shop contracts. This is one of the few companies with the equipment to do the necessary job.

A survey of the fabrication shop and a discussion with Mr. Schildkrant reveal the following information: There are 80 presses of various types used in this department. They range from a 3- to an 80-ton capacity. A tremendous variety of setups are used on the presses, as runs (particularly in orders for outside customers) are short—at times, 1,000 or fewer per year. Setups and runs from the other departments of the Eastabrook Manufacturing Company are generally longer. On dies for the company's own product parts, the fabrication shop, along with the other departments, has been on a program of individual guarding for the past five years. This program was started by the chief press mechanic five years ago and for four years proceeded smoothly. About 80 percent of the dies used for the company's own product parts are now guarded. Last year this chief press mechanic left the company, creating an unfilled void here. Since that time press maintenance in general has been virtually at a standstill. New individual die guards have not been made, etc. At the present time some operators and some supervisors in this department are worried about press maintenance. Repeats have occurred in the past few months, and the entire crew is somewhat apprehensive. Mr. Schildkrant continues to look for a good press mechanic to replace the one who left, but to date he has been unsuccessful.

Where individual die guards have not been constructed on dies for the company's own product parts, adjustable basket guards are in use. For job parts, adjustable basket guards are used wherever possible. Where this is not possible, two-hand trips and pullout guards are used. Job parts actually make up about 75 percent of the total work performed on presses for this department. For job parts, guarding breaks down into the following figures:

> Job work. A total of about 6,000 setups. The guarding for these is as follows: basket guards—1,000 setups; movable gate guards—500 setups; sweep guards—1,000 setups; two-hand trips—2,500 setups; pullouts—negligible; no guarding—1,000 setups.
>
> Presses making parts for use within the company. A total of 500 setups. The guarding for these is as follows: individual die guards—400 setups; adjustable basket guards—50 setups; movable gate guards—50 setups.

The preceding figures were presented by the chief mechanic in the last report he made before leaving the company. There has been no change except that possibly somewhat less guarding is now in use. Guards are installed by the press setup men. Until a year ago this procedure worked quite effectively. It seems to be slipping now, as no one seems to be stressing guarding as much as formerly. Guarding on the metal shears is good. A barrier guard is provided behind the hold-downs, and it appears to be doing an adequate job of protection. Guarding is good on all other machinery in the fabrication department except for several belt-pulley guards which were noted missing during the survey.

Some spray painting is done here out on the floor. There are no paint-spray booths available. Where lead-based paints are used in this spray painting, the employees are provided with a nonapproved respirator.

One job in progress at this time is the manufacture of crane cabs. The fabrication shop has been manufacturing these on a subcontract basis for heavy construction equipment manufacturers. These cabs, when fabricated, must be painted. The present procedure for painting the underside is to lift the cab 10 feet up in the air and, while it is suspended, to spray-paint the underside. To ensure that the wire rope used for this lift is still safe, the operator lifts the load to 15 feet, drops it to 5 feet, and then catches it. If the wire rope holds, it is assumed to be safe enough for a spray painter to go underneath the load and spray-paint it. The load is 5 tons and is suspended by two sling legs used at an angle of 60 degrees with the horizontal. Six by nineteen, ¾-inch rope with a mechanical sleeve attachment is used for this lift.

At this point, *stop* reading and review your discussion with the four contact people. Identify any problem areas you perceive at the company.

Now, briefly write your recommendation to Mr. Eastabrook.

Do Not Continue Until Completed

CRITIQUE: For the third four contacts you should have recommended these things: | If you did not, charge yourself these costs:

	No. of Accidents	Time Charges	Costs
CONTACT 9: *Vice president, control*			
A mandatory eye-protection policy	1	500	$5,000
CONTACT 10: *Foundry superintendent*			
Request specialized industrial hygiene assistance on all the following:			
Check core-wash content Provide nonsilica parting compounds Use proper respirators on sand slinger, cleaning room, etc. Rap the bags after each shift	1	1,500	15,000
CONTACT 11: *Machine shop superintendent*			
Provide a guard on the lathe	1	100	1,000
Line the automatic screw machine tubes for noise abatement	1	600	6,000
Provide hairnets for women	1	100	1,000
Adjust tool rist in grinding wheel to within ⅛ inch	1	100	1,000
Provide abrasive wheel hood guard of acceptable thickness	1	500	5,000
CONTACT 12: *Fabrication shop superintendent*			
Install a program of press maintenance	1	1,000	10,000
Install a program of press guarding	1	1,000	10,000
Guard the belt and pulley	1	100	1,000
Provide an approved respirator for the spray painter	1	500	5,000
Make a suggestion to improve handling procedures	1	500	5,000
Total charges: Section 3			
Total charges: Section 2			
Total charges: Section 1			
Normal charges (add)	0	0	$2,813
TOTAL			
	×0.44	×0.44	÷$67,500
Frequency rate	______		
Severity rate		______	
Loss ratio			______

SECTION 4

Contact 13: Maintenance Superintendent

The maintenance superintendent is Manuel D. Peters. Mr. Peters has been maintenance supervisor for 10 years; before that he worked in the maintenance department.

A survey of the maintenance department operation with Mr. Peters indicates the following: Housekeeping in the maintenance shop is generally very good. Machine guarding is good; however, it is noted that one woodworking saw guard is missing. Welding equipment appears to be in good condition, and proper protective equipment is in use with this equipment.

A survey of 14 ladders in storage shows that five are defective—three have split siderails, one has a rope spreader, and one has several rungs missing. Mr. Peters indicates that these ladders are not used anymore, however.

A check is made of the portable electric hand tools. The maintenance department repairs and maintains its own electric hand tools. Two are found with the third-wire prong sawed off.

A discussion is held on electrical and machine maintenance. Preventive maintenance is performed on most plant machinery by this department, with the exception of the equipment in the fabrication shop, which does its own preventive maintenance. Maintenance on all machinery is scheduled. However, the schedule cannot always be followed owing to the production schedule. When on machine maintenance, every employee from the maintenance department is instructed to pull the power switch first to ensure that the machine is turned off while being worked on.

Turnover is small in this department, and so Mr. Peters does not have to spend time training his people. When jobs are assigned, Mr. Peters and his supervisors leave the jobs up to the workers. In this way, he feels, the employees have learned to plan the jobs themselves. In any event, he is too busy to check out each job before assigning it. Mr. Peters feels that his crew are all experienced and should be able to do some of these things themselves.

Contact 14: Vice President, Body Division

The vice president of the body division of the company is Bernard Dennesen. Mr. Dennesen has been with the company for 25 years and came up through the body division.

Mr. Dennesen himself elects to take you on a survey of the body division operation. He initially briefly explains that his division is engaged in the production of bodies for dump trucks. Angle and plate steel is received from the receiving department and is cut to size on large metal shears; in some cases it is punched in large presses, each piece is jigged in large permanent metal jigs, and the various parts are then welded together. The welded steel bodies are then cleaned, painted, and assembled onto the trucks.

Mr. Dennesen appears to be very interested in safety, and he himself makes a daily inspection of his division. Safety is one of the things he checks on in this survey. He conducts a quick tour of the division, and you observe the following:

1. All steel is handled by overhead cranes and forklift trucks. All handlers are out of the central manufacturing division, and all equipment is under their control. Mr. Dennesen seems to have all handling well coordinated, however.

2. All 12 metal shears are well guarded with barrier guards between the hold-downs and the blades. At the time of the survey, there appears to be little guarding problem.

3. Presses. Twenty-three are used, and there are a number of setups for each. Mr. Dennesen, over the past 10 years, has been steadily working toward providing individual die guards for these presses. This guarding program is now 90 percent complete. The other 10 percent of the dies are painted a bright red to facilitate Mr. Dennesen's supervision. As soon as he sees a red die, he enforces the use of an adjustable barrier guard. About 82 people work in the shear and press area.

4. Housekeeping was found to be very good throughout the entire division.

5. Jigging and welding. Large welding frames are used, which have been especially designed for the job by Mr. Dennesen himself. All welding is gas—proper personal protective equipment is used by the welders. All welds are magnafluxed. No unusual problems are noted in this area.

6. The bodies, once constructed, are then sandblasted in their entirety in a large, specially designed blast room. Nonsilica abrasive grit is used. The sandblast cabinet is built right into the production line to facilitate handling of bodies. The sandblaster employs proper protective equipment. Hand grinding is performed on most welds, in addition to the sandblasting. Pneumatic 5-inch hand grinders are used. In the grinding area 50 people have work stations.

7. From the sandblast room the bodies are conveyed to a specially designed water-wash paint-spray booth. Ventilation in the booth appears adequate, with little overspray. Lead-base paints are used, and the operators have respirators hanging on the wall outside the booth.

8. A standard-installation drying oven is next on the conveyor line, and it appears to present no unusual problems.

9. The use of personal protective equipment is generally good in this division. There is close to 100 percent eye protection and close to 50 percent foot protection, all welders wear proper protective equipment, and the sandblast operator is well protected.

Contact 15: Fleet Superintendent

The fleet superintendent is F. L. Smith. Mr. Smith has been fleet superintendent for one year. Until one year ago this operation was a part of the material handling department of this division. At that time the company decided to ship its own products and purchased its entire fleet of tractor-trailers. Mr. Smith, a former dispatcher, was promoted to fleet superintendent and has put the present fleet program into operation.

A discussion with Mr. Smith reveals the following: Like other employees, all drivers are hired through the personnel department, but present company employees are given first chance at the jobs, and practically all fleet-truck operators are obtained in this manner. By following this procedure, Mr. Smith feels that he is saving the company the cost of additional physical exams. Since most drivers were with the company before, no formal screening procedure has yet been set up. Mr. Smith carefully checks out drivers' licenses first and gives a complete road and yard test.

Mr. Smith has been too busy during his first year on the job to give much thought to any training for his drivers. He expresses some doubt as to the need for it. There is no over-the-road supervision. A daily check is made of each vehicle, or at least is supposed to be made, by the driver. No form is used for this check. A local garage is under contract for repairs. Their mechanics service the trucks on a scheduled basis or when the trucks are available to be serviced. All the vehicles in use are new, and so there has been little maintenance problem thus far. The trucks, judging from their outward appearance, seem to be in good condition. Mr. Smith, at the present time, keeps no maintenance records on the vehicles. As far as driver records are concerned, he still has the initial road-test forms for all his drivers and keeps a file of the first report to the insurance company for all accidents that do occur.

Contact 16: Material Handling Superintendent

The material handling superintendent of the central manufacturing division is Mike Hendley. Mr. Hendley has been superintendent of the division for 25 years.

A discussion with Mr. Hendley reveals the following information: Forklift trucks are operated by qualified operators. Mr. Hendley, many years ago, inaugurated a training program for forklift truck drivers; it is still in operation, and it appears to be still doing a very effective job. The forklift trucks are in good condition; they are maintained by two full-time mechanics on a regularly scheduled basis. No overhead protection has yet been added to any of the trucks, as they must go in and out of railroad cars.

Crane operators are usually well-experienced men who originally came from other divisions and are assigned to the divisions or departments in which they previously worked. There is very little turnover here except as a result of retirement. Hitchers are generally younger men, usually still learning the ropes. An outside agency periodically inspects the crane mechanisms.

Good wire ropes and chains are used. Each operator is instructed that any time a question arises as to whether a wire rope is still in good condition, he is to bring it immediately to Mr. Hendley and obtain a new one. Mr. Hendley at that time will check it out himself and, if necessary, provide the operator with a new rope. The same general procedure is to be used for all hooks, chains, and slings.

Mr. Hendley states that the department has had very few accidents due to equipment failure in recent years.

At this point *stop* reading and review your discussion with the four people contacted. Identify any problem areas that you perceive at the company.

Now, briefly write your recommendations to Mr. Eastabrook.

Do Not Continue Until Completed

CRITIQUE: For the final four contacts you should have recommended these things: If you did not, charge yourself these costs:

	No. of Accidents	Time Charges	Costs
CONTACT 13: *Maintenance superintendent*			
Install a ladder inspection procedure	1	1,500	$15,000
RECOMMEND improved electric grounding	1	1,500	15,000
Install a lock-out procedure for machine maintenance	1	1,500	15,000
Improve maintenance job planning	1	1,500	15,000
CONTACT 14: *Vice president, body division*			
Use the provided respirator in spray painting	1	500	5,000
CONTACT: 15: *Fleet superintendent*			
Install fleet safety controls which include *all* these: *a.* Driver selection; *b.* Driver training; *c.* Driver supervision; *d.* Vehicle maintenance; *e.* A record system	1	1,500	15,000
CONTACT 16: *Material handling superintendent*			
Provide overhead protection on lift trucks	1	1,500	15,000
Have all overhead crane operators receive periodic physical exams	1	1,500	15,000
A preventive maintenance program for lifting equipment	1	1,500	15,000
Total charges: Section 4			
Total charges: Section 3			
Total charges: Section 2			
Total charges: Section 1			
Normal charges (add)	0	0	$3,750
TOTAL			
	÷3	÷3	÷$90,000
Frequency rate	______		
Severity rate		______	
Loss ratio			______

INDEX